ENVIRONMENTAL BIOLOGY

ENVIRONMENTAL BIOLOGY

Sushila Bhatnagar

ANMOL PUBLICATIONS PVT. LTD.
NEW DELHI-110 002 (INDIA)

ANMOL PUBLICATIONS PVT. LTD.
Regd. Office: 4360/4, Ansari Road, Daryaganj,
New Delhi-110002 (India)
Tel.: 23278000, 23261597, 23286875, 23255577
Fax: 91-11-23280289
Email: anmolpub@gmail.com
Visit us at: www.anmolpublications.com

Branch Office: No. 1015, Ist Main Road, BSK IIIrd Stage
IIIrd Phase, IIIrd Block, Bangalore-560 085 (India)
Tel.: 080-41723429 • Fax: 080-26723604
Email: anmolpublicationsbangalore@gmail.com

Environmental Biology

First Edition, 2010

ISBN 978-81-261-4527-0

PRINTED IN INDIA

Printed at Balaji Offset, Delhi.

Contents

Preface

Environmental Biology deals extensively with the elements, components, natural resources, physical, chemical, geological and biological aspects of environment. The last thirty years have witnessed a vast expansion of both interest and knowledge in environmental sciences. There has been a widening public awareness of the importance of environmental interactions with particular focus on such issues as the impact of pesticides upon food chains and the loss of global biodiversity owing to habitat destruction.

All the living beings, the animals and the plants should not be studied separately owing to their almost simultaneous origin. Both of them have shared same environment for their intensification and propagation. The surroundings may be natural, physical, chemical or biotic.

The study of plants, and their relationships with their surroundings is gaining more popularity with the term "Environmental Biology". Nevertheless, both the terms are synonymous. It would not be out of place to mention that the environmental biology is not a new discipline but simply an extension of ecological approach which stresses the study of the environment in its totality with special emphasis on the welfare and comfort of man who has been considered the most evolved and an intelligent unit of the ecosystem.

Author

Chapter 1

Nature of Water

It is truistic but nonetheless meaningful to say that the most important single feature of lakes and streams is water. Water is a biological necessity of plants and animals. It is required by plants for photosynthesis and by both plants and animals in general metabolism. Water in the lake basin or stream channel is of basic import ecologically, for it is the medium in which the members of the community live.

This is to say that the inhabitants must be adapted for finding shelter and food, utilizing dissolved gases, and reproducing in an environment very different from the terrestrial one. As an environmental medium, water enters into and maintains the total economy of the ecosystem. Therefore, before considering the physical and chemical features of inland waters and estuaries it seems proper to give brief attention to some of the properties of water as such. These attributes are basic to the environmental processes and characteristics of waters in lakes, streams, and estuaries.

THE WATER MOLECULE

The water molecule is constructed from covalent bonding of two atoms of hydrogen (H) and one atom of oxygen (O). In combination these give the familiar H_2O. Two chemical bonds unite the two H atoms to one O atom in the form H-O-H. The nucleus of each atom is positively charged; negative electrons revolve, or orbit, around the central core. The path of an electron is called an *orbital*, or energy level, and the bonds between a hydrogen and oxygen atom are formed by the overlapping of the orbitals in a fashion such that each atom contributes one electron which, in effect, completes the orbital of both atoms. Four of the unshared electrons of the oxygen atom, important in some reactions, are located considerably farther from the nucleus than are the remaining two. Because of the greater positive charge on the oxygen nucleus, the bonding electrons are not distributed equally between the hydrogen and oxygen nuclei, but are relatively closer to the oxygen atom. This results in a partial positive charge on the hydrogen atoms and a partial negative charge on the oxygen atom, or polarity.

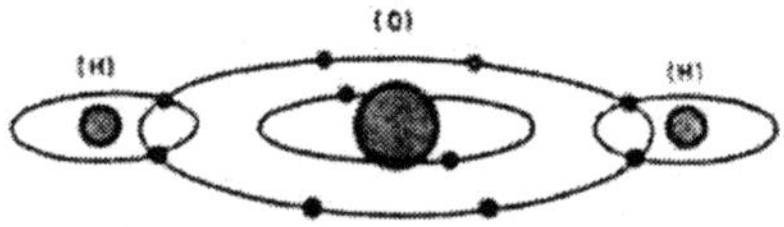

Fig. Diagram of the Water Molecule Showing the Overlapping Orbitals and the Shared Electrons.

Water is a highly organized liquid. The molecules are arranged in a definite configuration. The association of the molecules by hydrogen bonds into chains giving liquid water or ice, and the dissociation of the molecules to form vapour, account, in large part, for the versatility of water. Chemical reactions such as oxidation and hydrolysis occur when chemical bonds between the hydrogen and oxygen are broken. Physical processes, such as the melting of ice and evaporation, take place when the attractive forces between water molecules are broken without affecting the molecule.

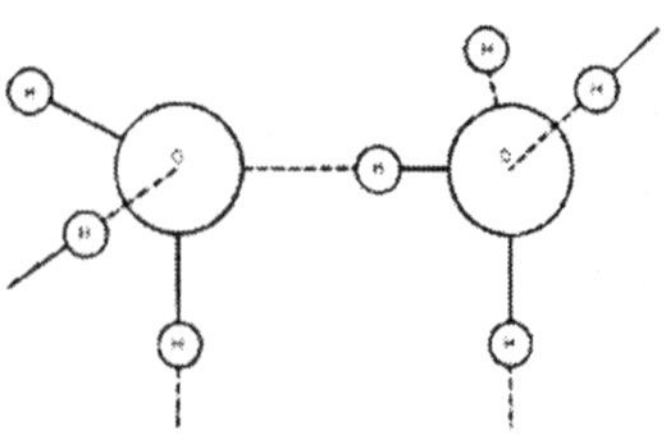

Fig. Two Molecules of Water Joined by Hydrogen Bonding. Chemical bonds uniting two hydrogen atoms with one oxygen atom are shown as solid lines. Hydrogen bonds are represented by dashed lines. This is but a portion of a vast system of variously oriented molecules in liquid water.

Ordinary water is not merely an aggregation of individual water molecules composed of two hydrogens of atomic weight 1, and an oxygen of atomic weight 16. Water also contains isotopic substances such as "heavy water" in which two deuterium (H^2) atoms are combined with an oxygen. There are also lesser amounts of isotopic molecules in which O^{17} and O^{18} replace O^{16}, and very small quantities of tritium (H^3).

PHYSICAL AND CHEMICAL FEATURES OF PURE WATER

HEAT OF VAPORIZATION

Compared with other liquid compounds of similar, simple molecular composition, water vaporizes very slowly when heated, i.e., water has a very high latent heat of vaporization (evaporation). Vaporization is a process in which thermal energy is supplied to overcome the attractive forces between water molecules (mainly hydrogen bonds in liquid water). A given volume of liquid water contains a larger number of hydrogen bonds per unit volume than most other common solutions.

We have just seen that there are two hydrogen bonds linking each water molecule to its neighbors; therefore the heat required for evaporation is approximately twice the hydrogen bond energy (4.85 kcal per molecular number of H bonds, or 9.7 kcal per gram molecular weight). Depending upon temperature, the amount of heat required for the vaporization of 1 g of water ranges from about 500 to 600 cal (9700 cal per gram molecular weight). In other words, as much heat is used to vaporize 1 g of water as to raise the temperature of 540 g by 1°.

SPECIFIC HEAT

The specific heat of a substance refers to its capacity to absorb thermal energy in relation to temperature change at constant volume. Water holds a great amount of heat with a relatively small change in temperature. The unit of measure of specific heat is the gram-calorie, or small calorie, which is defined as the amount of heat required to raise the temperature of 1 g of water from 14.5° to 15.5°C. Thus, water is the basis of this parameter and is said to have a specific heat of 1; the specific heat of other compounds is calculated as the ratio of their heat capacity to that of water. Only a few substances, including lithium at high temperatures and ammonia, have a higher specific heat than that of water.

This heat capacity of water has far-reaching implications. For one thing, it acts as a buffer against wide fluctuations in temperatures, thereby ameliorating terrestrial climates near large bodies of water. Secondly, an aquatic organism is subjected to much narrower ranges of temperatures than a land form: whereas land areas may reach 38°C or more, lakes seldom exceed 27°C. Furthermore, the slow rate of seasonal cooling and warming of lakes and streams is attributable to the high specific heat of water. For the same reason, changes in lake temperatures lag behind atmospheric fluctuations. In natural waters, dissolved substances lower the specific heat.

Much heat is necessary to bring water to boiling because a considerable amount of heat is dissipated in energy of vaporization before the boiling point is reached. The "reluctance" of water molecules to separate and emerge as vapour is due, as we have seen, to the hydrogen bonding of the molecules. Methane (CH_4) has nearly the same molecular weight (16) as water (18) but boils at -161°C as compared with 100°C for water. The difference is found in the fact that the outer electrons of the methane molecule are bonded chemically, there being no free pairs of electrons to enter into hydrogen bonding as in water.

LATENT HEAT OF FUSION

The heat taken up in the change of water from a solid state to a liquid phase with no change in temperature is termed the latent heat of fusion. Specifically, 79.7 cal of heat are required to melt 1 g of ice at 0°C. This is about 15 per cent of the heat necessary to separate the hydrogen bonds in the vaporization process. As heat is applied to ice, the molecules are set into motion.

Increased motion parts the hydrogen bonds until finally the molecular lattice collapses and brings about melting. There is evidence that some few bonds may be broken before melting begins, but even so, only about 15 per cent of the remainder is separated in the actual melting of ice; the fission of the remaining bonds accounts for the high specific heat and heat of vaporization of water.

DENSITY QUALITIES

Most liquids, including water, contract and become heavier with cooling: water is less dense, or lighter, at high temperatures, and becomes more dense as temperature is decreased. Fresh water is unique, however, in that it reaches its maximum density not, as with other liquids, just at the freezing point, but rather at 4°C. As it cools below this point the density decreases, that is, the water again becomes lighter as the crystalline structure develops. Upon freezing, water adds approximately 1/11 to its liquid mass. Thus water pipes burst and ice floats.

Water is one of very few substances whose solid state is lighter than the liquid. This quality is important limnologically, for it accounts for the fact that lakes freeze from the surface downward and that only small bodies of water, and lakes in the coldest regions, freeze solidly in winter; thus in most lakes and streams life goes on underneath the ice—albeit at a reduced rate.

Density is not only temperature-related, it is also a function of pressure, and of the concentration of substances dissolved or suspended in the water. High concentrations of dissolved salts account for increased density of ocean waters and the waters of unique types of lakes such as Great Salt Lake. Organic material such as finely divided detritus and organic silt, being heavier than water, will, in sufficient quantities, increase the density. Similarly, clays and fine muds increase the density of waters in lakes and streams.

Viscosity is a measure of the internal, molecular friction of a liquid and is concerned with mobility and flow. In order to flow, the bonding forces holding molecules of a liquid in form must be overcome, and there must be a void in the lattice pattern into which a loosened molecule can move. Compared to many liquids, gasoline for example, water is highly resistant to flow. This resistance is due to the relatively great energy contained in the hydrogen bonds of water molecules.

This quality of water limits stream discharge, and also relates to living habits, morphology, and energy expenditure of aquatic animals inhabiting a dense environment which offers considerable hindrance to movement.

From what has been shown of the activity of molecules in relation to temperature, we should expect viscosity to diminish with increase in temperature. Actually, viscosity changes about 3.5 per cent with each degree of temperature, although the relationship is not uniform.

Mass, or weight, of water is also related to density. By general standards, water has considerable weight. Although it varies with temperature, the weight of 1 cu ft of water is generally given as 62.4 lb (998.4 kg/m^3). If determinations are made below the surface, atmospheric pressure must also be considered. At a depth of 100 ft (30.5 m) the pressure is approximatey 58 lb per sq in. (4 kg/cm^2) or about 4 atms (pressure of water increases approximately 1 atm per 10 m depth).

ADHESION, COHESION, AND SURFACE TENSION

Adhesion is the tendency of a liquid to cling to the surfaces of some materials by means of bonds established between the hydrogen atoms of the water molecule and oxygen atoms of the other substance. Cellulose in wood, for example, contains a great number of oxygen atoms and is, therefore, very wettable.

Cohesion is the property of liquids which offers resistance to being pulled apart or to the formation of new surfaces. It has been calculated that a force of 210,000 lb would be required to break a column of pure, perfectly formed water of 1 sq in. (6.45 cm^2) cross section.

At the surface of a water mass the force mainly responsible for such resistance is termed *surface tension* and results from the unsymmetrical activity of water molecules at and below the surface.

Whereas the internal molecules are bonded on both sides, those at the surface are attached only to the molecules below, there being none above. A force is thus imparted inwardly giving rise to a surface phenomenon analogous to a taut sheet over the water.

The surface tension at the zone of contact between air and water is 73.5 dynes at 15°C. Surface tension varies inversely as the water temperature, and is lowered by the presence of organic substances. The addition of inorganic salts tends to increase surface tension.

SOLVENT ACTION

The phenomenon whereby substances may be dissolved and retained in solution is exhibited to a greater extent by water than by any other liquid in nature. Of the known chemical elements, about 50 per cent have been reported in water and it seems probable that at least traces of all elements may occur. This property is, of course, of prime importance to the maintenance of life in the aquatic environment. Basically two types of processes involving hydrogen bonds are concerned in the dissolving action of water.

One type of solvent process may be described as *inert*, for, by means of hydrogen bonding, the dissolved substances are relatively unaffected by the solvent. Many compounds of ammonia, nitrate, and phosphate, as well as sugars, alcohols, and various organic acids, are bonded to the hydrogen atom of the water molecule through oxygen atoms, hydroxyl groups, or through nitrogen bound to hydrogen.

It should be pointed out that these compounds are some of the important ones involved in energy transfer and storage in biological systems. Therefore, through hydrogen bonding the substances are delivered unmodified and readily accessible, not only from environment to plant or animal, but within the physiological systems of the organisms as well.

The second type of solvent process involves the separation of the electric charge between the hydrogen and oxygen atoms in the H-O-H molecule. Water is characterized by a very high charge separation; thus various salts such as sodium chloride and the salts of potassium are retained in solution.

In still another activity, water has a remarkable ability to ionize dissolved compounds through the separation of the bonds in the H-O-H molecule. This process is due to the high dielectric constant of water. The molecule shows polarity, and total separation of one H from OH results in two charged ions, H+ and OH-. The H+ becomes the hydrogen ion of acids and the OH- becomes the hydroxyl radical of bases. A simple, reversible ionization reaction is that involving the solution of carbon dioxide in water:

$$C0_2 + H_2O \rightarrow H_2CO_3,$$

or more completely:

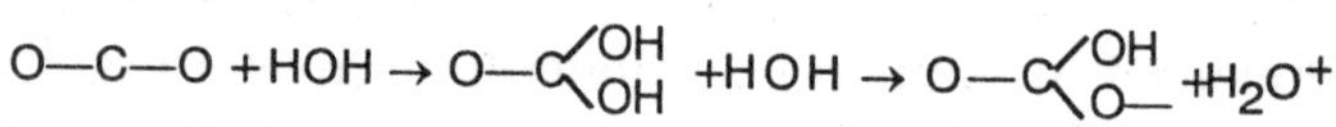

TRANSPARENCY

Pure water is quite transparent. Since the extent of penetration and absorption of light-wave components, so important in plant and animal relationships, is a function of transparency, this characteristic of water is of great interest to limnology. Pure water absorbs solar radiation selectively, the major absorption being at the red end of the spectrum. Minimum absorption is in the blue range at about 4700 Angstrom units (A). Approximately 90 per cent of the radiation above 7500 A is absorbed by I m of water.

LIQUID NATURE

The attribute of being liquid constitutes one of the more unique properties of water, for liquids in nature are not common. In addition to water, only one other inorganic substance occurs in liquid form at the earth's crust; this is mercury. It has been suggested that liquid carbon dioxide may be found in quartz, but this is apparently not fully confirmed. Petroleums are liquids, but are of organic origin.

Chapter 2

Solar Radiation and Natural Waters

At this point it might be well to refocus our attention on an idea suggested in the Introduction to our study of inland waters and estuaries. Recall that there appear to be three major themes running through our approach to the ecosystem:

- Fitness of the environment,
- Evolutionary adaptations of the living components of the systems,
- Traffic in energy.

Of the three, two are more or less directly dependent upon solar radiation in order to maintain their importance and to fulfill their roles in the ecosystem; we refer specifically to the nature of the environment and to energy transfer. As we shall presently see, temperature, as a characteristic of natural waters, is of paramount importance in regulating many chemical, physical, and biological reactions; it is, of course, derived mainly from solar radiation. Green plants constitute the major source of energy-containing substances available to animals in natural waters. From this relationship stem the intricate food webs involving energy traffic; the ultimate source of this energy is solar radiation.

During the course of a year the amount of radiant energy received by the earth from the sun approaches the staggering figure of $1._3 \times 10^{21}$ kcal. Some of the solar radiation is reflected from the earth's surface by living features such as vegetation, and by nonliving components represented by land forms, lakes, and seas. A portion of the spectral array is utilized by green plants in synthesizing energy-containing substances necessary for the sustenance of plant and animal life. In addition, heat enters into the development of meteorological processes such as wind and rain which, in themselves, create new habitats and modify old ones.

The known range of radiations in the electromagnetic spectrum extends from extremely short-wave lengths of gamma rays of the order of 0.001 Å (1 A = 1 × 10^{-8} em) to exceedingly long Hertzian rays of 3 × 10^{14} Å. Much of this total radiation is distributed outside the earth's surface. The direct solar radiations that impinge on the surface of the earth range from about 135,000 to 2860 A; the most intense radiation, however, extends between 3000 and 13,000 Å. The peak of radiation distribution is in the bluegreen range, about 5500 Å. Of those radiations that arrive at the water surfaces on the earth, the infrared components (about 0.1 mm to 7700 A) contribute largely to heating. Within a narrow segment of the spectrum lie rays of the so-called light, or visible, spectrum.

These range from 7700 to 4000 A. Below the visible spectrum are found the components of the ultraviolet segment of the solar spectrum, the range being from 4000 to about 2860 A. The earth also radiates, the emissions falling in the infrared range of long waves because the earth is a cool body. The earth's radiations are not entirely lost to outer space because of absorption of some of the radiation by moisture in the earth's atmosphere. At the same time, the atmosphere differentially absorbs solar radiation (on the order of 15 per cent). The net result is the retention of a great portion of the earth's radiation and the transmission of about 85 per cent of solar radiation by atmospheric water vapour.

The phenomenon whereby the earth's atmospheric water vapour transmits most of the radiant energy of sunlight while absorbing the radiant energy of the earth, has been called the "greenhouse effect." The principle of the botanical greenhouse is based on the fact that glass of the sides and roof permits the passage of solar energy, but inhibits the transmission of a great part of the "black body" energy given off from the plants and other structures inside the greenhouse. In other words, the contents of the greenhouse absorb solar energy which passes through the glass; the contents in turn radiate an energy of longer wave length which is not transmitted through the glass. This long-wave heat is therefore retained inside the greenhouse. Actually, greenhouse thermodynamics is more complicated than here indicated, but this simplified statement should point up the analogous roles of greenhouse glass and the layer of atmospheric water vapour in retaining heat.

Of interest to us in terms of light available for photosynthesis and for heat in maintaining the environment is the mean value of solar radiation falling upon the earth's surface. It has been calculated that an average of approximately 1.5 g-cal/cm^2/min of radiation is incident upon the earth at sea levcl. This value is to be compared with the so-called solar constant of 1.92 g-cal/cm^2/min reckoned as the radiation received on the outer surfaces of the earth's atmosphere.

Actually, the light falling upon any area of the surface of the earth may be composed of direct sunlight and scattered light from the sky, or indirect solar radiation. Both of these are subject to considerable variation. The magnitude of direct solar radiation striking a given point depends upon season, geographical location with respect to latitude, time of day, the angular height of the sun and elevation of the point under observation, and the transmission quality of the atmosphere. The quantity of indirect solar radiation received at a given point is quite variable, but generally constitutes about 20 per cent of the total radiation.

All the light received at the surface of a body of water does not enter the water. A portion of the radiation is reflected, that amount being a function of the angle at which the light strikes the surface of the water and the condition of the surface. From an undisturbed surface, the greater the angle of incidence from the perpendicular, the greater the reflection.

When the light rays strike the surface at a very low angle, as much as 35 per cent of the light may be reflected. At high angles the reflectivity is on the order of 5 to 10 per cent. We are concerned limnologically with the light that penetrates the water surface.

One of the conspicuous features of pure water is its high transparency. Pure water does not exist in nature, however; yet the penetration of light into water, and the fate of the spectral components within water, are important in ecosystem relationships, and therefore of primary importance to students of aquatic communities.

The deviation of transparency of natural waters from pure, laboratory water ranges widely, both quantitatively and qualitatively, being determined primarily by dissolved substances, suspended materials, organisms, latitude, season, and the angle and intensity of the entering light. In order to more fully appreciate optical phenomena in, and optical qualities of, natural water, let us first consider some principles derived from laboratory studies of pure water.

The total light entering and passing through a given column of pure water undergoes a reduction in total intensity and change in spectral composition with depth. These processes of reduction and change result from scattering and differential absorption of light by water.

The reduction in intensity, or quenching, of the light entering the surface is expressed in the following relationships based upon Lambert's Law:

$$I_z = I_o e^{-\eta z} \quad \frac{I_z}{I_o} = e^{-\eta z}$$

in which Io is the original intensity, of the entering light, Iz the measurement of intensity incident at depth z, and e is the basis of natural logarithms; n becomes a constant for a particular wave length, known as the *extinction coefficient*, or the percentage of the original light held back at depth z. The percentage of light passing through z is termed the *transmission coefficient*.

These equations are based on the use of pure water and monochromatic light. Since neither of the conditions occurs in nature, factors which further reduce transmission through the distance Io - 1z must be applied for natural waters. Such factors would include, in addition to water, suspended materials and dissolved substances.

It has been demonstrated that in pure water at a depth of 70 m the original intensity of blue light is reduced some 70 per cent, and the yellow component shows only 6 per cent transmission beyond 70 m. The red component is not transmitted beyond 4 m, and orange is quenched in about 17 m. In general, approximately 53 per cent of the total incident light is transformed into heat and undergoes extinction in the first meter of water. The longer wave lengths (red and orange) and the shorter rays (ultraviolet and violet) are reduced more quickly than the middle-range wave lengths of blue, green, and yellow.

LIGHT IN LAKES

In natural waters, the blue segment is transmitted farther than all others. At depths beyond about 100 m blue light is the sole illumination. In highly colored or stained waters, orange and red are transmitted more deeply than other components, but all are rather quickly reduced. In many moderately transparent waters, the greatest transmission is in the yellow.

The most accurate and precise measurements of solar radiation in natural bodies of water are best made with photoelectric apparatus, employing various colour filters. A photocell contained in a waterproof compartment is connected with a galvanometer and the cell lowered to desired depths. Direct readings of under-water radiation falling upon the cell are made from the galvanometer at the surface.

An American, Professor E. A. Birge, pioneered in the use of electrical equipment for measuring light in lakes. This instrument, developed by Birge and physicists at the University of Wisconsin, was first used in 1912. Called by Birge a *pyrlimnometer*, it measured the total radiation which impinged upon the surface of a thermopile, the thermal current produced being read from a galvanometer.

In some ways, particularly with respect to uniform sensitivity and the expression of energy units, the thermopile remains unsurpassed. But it is costly, and thus the relatively inexpensive photocell has come to be used widely in limnological work. Welch explains in great detail this and other methods of light measurement.

About 1905, Birge was joined in Wisconsin by Chancey Juday and together their researches and reports gave impetus to limnology. The effects of several of the previously discussed factors which relate to transparency of natural waters and comparison with pure water are seen in studying this figure. Total solar radiation is shown as incoming sunlight by the circle with a shaded area at the bottom right. The invisible part of the spectrum, mostly infrared, is represented by shading and includes about half of the entire area.

In this, and the other circles, the portion representing the visible spectrum is divided into sectors proportional to the quantity of each of the six colors measured by Birge and Juday. Beginning at the heavy, horizontal line in the "three o'clock" position, observe that the segments are labeled red, orange, yellow-green, blue, and violet in order of decreasing wave length. The enclosure of the red, orange, and yellow segments in a heavy line, and the arrow at "nine o'clock" will be referred to later.

The circles representing pure water and the four lake examples at given depths are drawn to scale proportionate to the total light present at the depths indicated and to the circle representing incoming sunlight. The unabsorbed light remaining at each depth is given as a percentage outside the circle. Variations in each of the six components of the spectrum are indicated by, the area of the circle segments.

Immediately, apparent is the rate at which transmission decreases with depth, and the variations among the lakes. Similarly, great differences in colour changes are seen. It is interesting to note that the over-all rate of light extinction and the proportion of the colour segments for Crystal Lake are not too different from pure water. In both Crystal Lake and pure water, the yellow-orange-red segments decrease, and the green and blue increase with depth.

Accompanying this is a swing of the arrow in a counterclockwise direction. The data indicate that red waves are absorbed rapidly in the first 3 or 4 m and that with increasing depth the light becomes greenish-blue. In the series for pure water the proportions represent the absorption of direct sunlight; the lake determinations include mixed light from sun and sky. Since mixed light contains proportionately more blue, this light would be transmitted more effectively than sunlight alone in pure water.

As the diagrams for pure water and Crystal Lake would indicate, these waters have no stain. The figures for the other lakes, Silver, Midge, and Mary, indicate increasing amounts of stain and colour and a proportionate decrease in transparency and light transmission. There is also an increase in absorption and reduction of the green, blue, and violet components. In contrast to pure water and Crystal Lake, the arrow swings clockwise in these stained lakes, especially in Midge and Mary. The last two are bog lakes and colored dark brown by dissolved and suspended organic materials from the surrounding bogs. In these lakes, orange and brown are the conspicuous spectral components, but nearly all of the incoming light is absorbed in the upper 2 m.

SECCHI DISK TRANSPARENCY

A much older method of measuring transparency of water, and one that still holds considerable value, is the use of the Secchi disk. The method was devised by A. Secchi, an Italian, in 1865. In practice, a white disk, 20 cm in diameter, is lowered into the water by means of a line with measured intervals.

The arithmetical mean of the distance at which the disk disappears from view in descent and that at which it reappears in ascent is given as the Secchi disk transparency. In employing this device, care should be exerted to standardize techniques and conditions. Reflected light from the bottom will introduce an error since the technique involves comparison of the brightness of the disk with the bottom brightness. It has been calculated that the disk disappears at approximately the region of transmission of 5 per cent sunlight.

In very turbid lakes the Secchi disk transparency is quite low. In Lake Texoma, on the Texas-Oklahoma border, values on the order of a few centimeters occur in places. Similarly, artificial farm ponds of the southeastern United States often give readings of less than a meter, particularly, following fertilization when blooms of plankton appear. Secchi transparency values near 18 m are characteristic of certain lakes in the Convict Creek Basin of California. Generally, values above 30 m are unusual. Crater Lake, Oregon, however, has a transparency near 40 m.

COLOUR AND TURBIDITY OF LAKES

Colour and turbidity are here treated together because of their somewhat similar roles in giving various hues and other optical qualities to lakes; one distinction will be made, however. Both of these qualities determine light transmission in natural waters and consequently "regulate" biological processes within the bodies of water. To varying degrees, both give some qualitative indication of the productivity of the waters when simply viewed from above. The general chemical nature of lakes may often be deduced from the colour and turbidity of the water. Before proceeding further, we should define our terms in order to avoid confusion.

COLOUR

The colour that we perceive is made up of unabsorbed light rays remaining from the original entering light, but now passing out of the lake. Completely pure water should absorb all light components and appear nearly black. This is not seen in natural bodies of water, however, for lakes containing little suspended materials usually appear blue.

The blue hue is probably the result of scattering of light by water molecules in motion, the effect being proportional to the fourth power of the wave length of the light components. Since the scattering of light of long wave lengths is less than that of short, the blue predominates.Such hues as we normally see, however, may range from green-blue through blue-green, green, yellow, yellow-brown, to brown.

The vivid, poetic, and not necessarily imaginative impressions such as the "skyblue waters" of Cadman's song,

Thoreau's "pure sea-green Walden water," and the "silver" of Seneca Lake as seen by James Gates Percival, all suggest the rich variety of colour interpretations possible, depending upon one's point of view.Limnologically, *true colour* (sometimes called *specific colour*) of natural waters is derived from substances in solution or from materials in colloidal state.

Apparent colour, on the other hand, is usually the result of interplay of light on suspended particulate materials together with such factors as bottom or sky reflection. In order to realise true colour, a sample of the water should be filtered or centrifuged, thereby freeing the water of sources of apparent colour. The numbers in parentheses near the names of the lakes refer to the colour of the water in units of the *platinum-cobalt scale* of the United States Geological Survey.

The basic technique involves comparison of lake waters with a series of dilutions of a solution of potassium chloroplatinate (K_2PtCl_4) and crystalline cobaltous chloride ($CoCl_2 \cdot H_2O$). The units are called platinum-cobalt units, based upon 1 mg Pt/l as a standard, and range from zero in clear waters to over 300 units in very dark waters of bog communities.

The Geological Survey set of glass disks corresponding to the platinum-cobalt series is now rather widely used in place of the liquid dilutions. The major classes of dissolved and particulate matter found in typical lake waters may be stated as follows:

Dissolved Substances:

- Proteins and related compounds
- Fats and related compounds
- Carbohydrates and related compounds
- Break-down derivatives of all three of the above

Particulate organic matter (Seston):

- Living organisms (*Plankton*): mostly microscopic forms.
- Phytoplankton: plant types
- Zooplankton: animal types
- Nonliving particles (*Tripton*): dead organisms, detritus, colloidal substances

The average year-round percentage composition (by weight) of these categories in a "typical" lake is shown in Figure. For any particular lake the bar lengths would doubtless vary greatly; swamp or bog lakes would show a higher percentage of dissolved organic matter; alkaline lakes would be richer in dissolved salts.

The important point here, however, is that all elements of these classes, individually or in concert, contribute to the colour and turbidity of natural waters.

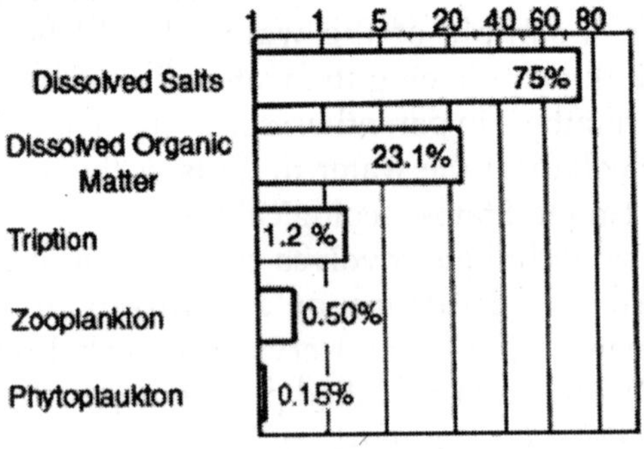

Fig. Year-round Average Percentage Composition of Sestonic Components and Dissolved Organic and Inorganic Material in a Lake.

For the most part, lake colors are determined by the predominating components of the *seston*, the mass of various living and nonliving substances in the water. Plankton algae are most frequently responsible for certain colors; an abundance of blue-green algae imparts a dark greenish hue; diatoms give a yellowish or yellow-brown colour.

Zooplankton, particularly certain of the microcrustaceans, may tint the water red. Humus often causes water to be green, or yellow-brown, the darkest brown coming from extractives of peat.

Suspended colloidal inorganic substances may account for certain tints. Calcium carbonate in lakes of limestone regions results in a greenish colour.

Volcanic lakes may be yellow-green from sulpher, or red from ferric hydroxide. Generally, a rich, highly productive lake may appear yellow or grayblue or brown due to quantities of organic matter. Less productive lakes tend toward blue or green. We have just seen that the colour of lake waters differs widely, depending upon the nature and quantity of dissolved and suspended materials, the quality of the light, and other factors.

It is worth noting also that colour within a given lake may not be uniform from surface to bottom. Studies in Weber Lake, Wisconsin, revealed that in August the epilimnion was transparent and contained but little organic material; the hypolimnion, on the other hand, was highly colored.

At that time, the surface water had a colour rating of 12 units, the lower limits of the metalimnion (about 8 m) 81, and at the bottom the colour had increased to 93 Pt-Co units. A similar phenomenon has been observed in a New Jersey stone quarry lake. It is suggested that the increase in colour in the depths of such waters is due to organic substances derived from bottom sediments and also to increased phytoplankton populations supported by such substances.

In a similar vein, the colour of lakes may change periodically. Seasonal increases in surface runoff contribute great quantities of inorganic and organic substances which, as we have seen, impart various colors. Summer or early autumn production of phytoplankton "blooms" causes a lake to become a "soupy green," which disappears later in the season. Exposure to light causes the bleaching of certain colors in natural waters. This process also results in variations in the colour of a given lake, such variations appearing seasonally with cyclic fluctuations in light intensity and angle of incidence. The bleaching effect should be felt vertically within the lake, depending upon transparency.

TURBIDITY

We have seen that colour in natural waters is attributable in great part to suspended materials of varying particle size and composition. *Turbidity*, on the other hand, is the term used to describe the degree of opaqueness produced in water by suspended particulate matter. While the nature of the materials contributing to the turbidity is mainly responsible for the colour quality, the concentration of the substances, if sufficiently high, determines the transparency of the water by limiting the light transmission within it.

The kinds of materials creating turbid conditions in a given body of water are as varied as the biotic and abiotic composition of the surrounding terrain, inflowing streams, and the lake itself. Substances such as various grades of humus, silt, organic detritus, colloidal matter, and plants and animals produced outside and brought into the lake, are termed *allochthonous*. Turbidity-creating matter produced within the lake is said to be *autochthonous*. Both contribute to the total quantity and quality of lake turbidity.

Although several methods for measuring turbidity have been devised, two are perhaps most widely used today. One technique involves the U. S. Geological Survey turbidity rod, which consists of a marked staff calibrated against known concentrations of a standard material (usually fuller's earth) with a length of platinum wire at one end and an eyepiece at the other. The end bearing the platinum wire is lowered into the water until the wire vanishes from view. The calibration mark at the level of the water surface gives the turbidity in parts per million (ppm).

Devices such as the Jackson Turbidimeter and the Hellige Turbidimeter make use of light penetration through a sample of water. In the former apparatus, the disappearance of the flame of a standard candle when viewed vertically through a column of water is a measure (ppm) of turbidity. The Hellige instrument compares a vertical beam of light through a column with the Tyndall effect produced by lateral lighting from the same source, usually an electric bulb. An adjustable aperture regulates the light reaching the eye of the observer and the aperture size is calibrated for turbidity and determined in parts per million. For details of these instruments refer to Welch.

Turbidity is not a uniform parameter even within a specific lake. Seasonal increases in stream discharge, for example, may introduce considerable amounts of silt and other sediments and materials, thereby altering the lake colour and turbidity. With decrease in stream discharge, much of the allochthonous matter begins to settle in the lake basin. The rate of settling is not the same for all classes of materials, grading from sand to colloidal particles.

In natural waters the settling process is complicated by certain attributes of water itself and by the dynamics of water movement, some of which have already been considered. In accordance with Stokes' equation, the rate of settling (velocity of fall of a spherical body through a liquid) is dependent upon gravity acceleration, the radius of the body, the viscosity of the liquid, and the specific gravity of the body and the liquid. Recall that viscosity is related to temperature and that temperature is not necessarily uniform throughout a lake, this being especially important during summer stratification. Turbulence and other water movements contribute to the rate of settling in lakes, particularly during spring and fall overturns.

Some classes of matter which exist in lakes do not settle under normal conditions. These include true colloidal systems or very fine particles of favorable specific gravity, and animal and plant forms capable of modifying their specific gravity and employing other mechanisms (flotation devices, locomotion) to maintain position. The particles in colloidal systems of silt are kept in suspension by the combined effects of turbulence and negative charge on the colloidal particles. It has been observed that the addition of acids to turbid ponds causes flocculation and precipitation of silt particles.

These processes are brought about by neutralization of the negatively charged particles by the positively charged hydrogen ions of the acids. The mechanisms and dynamics of settling and the relationships of particle size to distribution by water movements in lakes are not well known—this in spite of the obvious importance of turbidity in such processes as energy circulation of nutrients, and the effects of sedimentation in lakes.

One most important role of turbidity in the interrelationships of the aquatic ecosystem is the effect, previously mentioned, of suspended materials on transmission of light. This aspect has received considerable attention, especially as it relates to productivity and energy flow within the community. At this point we wish to consider only the effects of turbidity upon light; turbidity as a factor in productivity will be considered later.

The damping effect of suspended particles on the transmission of solar radiation, also termed "light-quenching," can be determined effectively from the use of the formula expressing Lambert-Beer's Law,

$$\frac{I_z}{I_o} = e^{-kdc}$$

in which the concentration (mg/l) of light-quenching material c, and the thickness of the column in which light is quenched d, are related to the ratio of observed light I_z to incident light I0 to determine the partial extinction coefficient *k* of suspended particles in natural waters. (This *k* is not to be confused with that representing over-all coefficient in some formulas.)

The parameter I_z may be measured with an underwater photometer. The value of c is the dry weight of suspended matter per unit volume of water after centrifuging. The advantage in the use of LambertBeer's expression over Lambert's Law lies in the recognition of variable *c*; in Lambert's law this factor is included in the extinction coefficient *k*.

The Lambert-Beer expression has been used to interpret data on light quenching in western Lake Erie and other lakes. In the Lake Erie investigation, I_0/I_z was set at 100 (observed light equals 1 per cent of surface light), and d represented the depth at the incidence of 1 per cent of surface light. During the study the highest value for c was found to be 35.9 mg/l with a *d* value of 1.2 m; thus k became 0.107. The highest observed value for *d* was 7.6 and for c was 5.4; a *k* value of 0.112 was then derived.

A significant datum was obtained in spring when it was found that the average depth associated with 1 per cent of the surface light was 3.5 m; this degree of turbidity represented a load of suspended matter near 10 mg/l. The rapid absorption of light and the quenching effect resulting from turbidity factors are vividly demonstrated.

It might appear that colour and turbidity act together to exert relative effects on light penetration. Such may not be the case in all instances, however.

Studies in Atwood Lake, a reservoir in Ohio, have indicated that colour and turbidity are very probably independent variables exhibiting no interaction with respect to transparency. It was further found that colour may be the major factor affecting light penetration, except of course during periods of introduction of large amounts of silt during heavy rainfall.

COLOUR AND TURBIDITY OF STREAMS

Basically, colour in stream waters derives from the same physical laws of light transmission, differential absorption by substances in the water, reflection, and back-radiation as operate in other natural waters. However, as a result of several factors streams generally fail to exhibit the great variety of colour that may be seen in lakes.

The upper reaches of most streams are characterized by clear waters, at least during the nonflood season. In these regions tile streams lack a true plankton which, as we have seen, may be responsible for certain apparent colors in lakes. Relatively "pure" shallow streams appear clear due, in part, to the fact that light is absorbed quite rapidly in the first meter or so.

In many areas, particularly in the southeastern United States, small, shallow streams which drain "flatwoods" and swamps are often colored a light to dark amber (about 160 on the U. S. Geological Survey Scale). This colour is probably attributable to dissolved plant substances such as tannin; the streams are typically acid. Similarly, certain small creeks in southern Indiana become 'inky black" in the autumn due to extractives from large accumulations of leaves in the water. Various extrinsic factors may also account for apparent colour in streams.

Rich growths of diatoms on rocks on the stream bed may lend to the water a brownish hue; algae may impart a greenish tint, and sulfur bacteria a yellow colour. Certainly pollution, from a variety of sources, can contribute to both true and apparent colour in streams and should not be disregarded in field investigations.

COLOUR AND TURBIDITY OF ESTUARIES

The causes and chemistry of colors in marine waters are not thoroughly known. It is likely, however, that dissolved substances impart certain hues or enter into colour shifts. For example, it is known that water-soluble yellow pigments are common in coastal areas and may contribute to the various shades of green in off-shore waters.

The blue of the sea results, as in inland waters, from the scattering of light by water molecules. Suspended detritus and living organisms give colors ranging from brown through red and green. The Red Sea derives its name from a brownish colour due to great numbers of particular algae.

As we move into coastal estuaries we find that the brilliant colors of the open seas are generally not apparent. In some estuaries of moderate tide and current action, blue and green hues are noted during the nonflooding periods.

Most estuaries, however, are characterized by dark colors resulting from typically high turbidity. On occasion various planktonic forms become so numerous that they give a reddish or greenish tint to the water. Dissolved materials such as tannic acid delivered by the inflowing stream of the estuary or from local decomposition of organic substances cause estuaries to be light brown.

The major component of turbidity in estuaries is, of course, silt. The volume of silt transported into estuaries by streams fluctuates seasonally, with the maximum discharge taking place during the wet season.

Some materials may be brought in from the sea, but these are usually minimal. In addition to allochthonous silt and other materials, much matter which contributes to turbidity originates from erosion within the estuary itself; the magnitude of autochthonous substances appears to be dependent upon the shape of the basin and prevailing currents. Throughout the year the amount of materials in suspension in an estuary decreases from the upper reaches to the mouth of the basin, that is, transparency increases downstream. This is due to diminution in velocity and carrying capacity of the inflowing stream current, and to the electrolytic effect of sea-water salts. The latter process involves the coagulation of negatively charged particles of colloidal silt by positive ions of certain metals present in sea water.

Even though there is typically a decrease in turbidity seaward along the axis of an estuary, its waters are decidedly more turbid than the sea. Evidence of this is seen in large areas of highly discolored water lying opposite the mouths of estuaries, contrasting vividly with the clearer sea water. Such masses of estuarine waters often extend many miles seaward or, depending upon currents, for great distances along the coast.

The major effects of high turbidity in estuaries are:

- The quenching of light penetration, thereby inhibiting photosynthesis and the production of plants,
- The building of deep zones of mud, silt, other sediments, and detritus.

In many estuaries, notably during periods of considerable stream inflow, light is reduced to 1 per cent of surface radiation at less than 3 m. In certain regions of the open sea the yellow-green components are not diminished to 1 per cent until a depth of nearly 100 m is reached. Transparency decreases shoreward such that in coastal waters the 1 per cent illumination level is generally between 15 and 30 m.

THERMAL RELATIONS IN FRESH AND ESTUARINE WATERS

From the board and basically ecological point of view, the thermal properties of water and the attending relationships are doubtless the most important factors in maintaining the fitness of water as an environment. We have already considered some of the physical properties which contribute to the total state of the hydroclimate.

Even so, it would be worth recalling the fact that the specific heat of water is among the highest of all substances.

This very great absorption capacity accounts for many conducive (and nonconducive) features of the aquatic environment. Remember also that water has the unique attribute of reaching its maximum density at 3.98°C rather than at freezing. These two characteristics will be important in considerations of thermal dynamics in fresh and estuarine waters.

At the beginning of the preceding section we learned of the range of solar radiation that impinges upon the earth's surface. We directed our attention mainly toward those components responsible for illumination so important to photosynthetic activity of plants.

At the same time we were aware of the presence of light of wave lengths longer than those in the visible range, namely the infrared rays which range from about 7000 A upward. We know from Lambert's Law that light absorption by water increases exponentially with the light path. In a column of water 1 m in length 91 per cent of light with a wave length of 8200 A is absorbed; only 9 per cent is transmitted through the column.

In 2 m some 99 per cent of this light is absorbed. Over 50 per cent of the solar radiation is absorbed within a depth of 2 m. It is this light with which we shall now be concerned, for the heating of natural waters results from this absorbed radiation. In other words, we should consider *temperature* as an intensity factor of heat energy.

TEMPERATURE AND HEAT IN LAKES

One of the most outstanding and biologically significant phenomena of lakes is found in the relationship between water and temperature as expressed by seasonal variations. In many lakes, these variations take the form of pronounced changes in the over-all thermal structure and dynamics.

During winter, the temperature of the water in moderately deep to deep lakes is relatively uniform from surface to bottom; or if ice forms, this colder layer floats on the underlying waters. In spring, circulation and mixing of water results, typically, in a uniform temperature from surface to bottom. During summer, the vertical distribution of temperature may come to resemble that shown in Figure.

In this case, the lake is essentially stratified. From the surface to about 15 m, temperature changes little with depth. Between 15 and 30 m, the temperature drops rapidly. The region of the lake below approximately 30 m is rather uniformly cool. In the fall, the summer condition is broken up by circulation and mixing similar to that of spring, resulting once more in uniform temperature of the water.

How do these conditions come about? The answers are found in light absorption, heat dynamics, density phenomena, and wind action. These are best appreciated by considering in more detail the *annual temperature* cycle of a lake.

SEASONS IN LAKES

In winter, as shown for March at the extreme left of Figure, ice at near 0°C covers the surface of the lake. Recall that ice floats because it has a density less than that of water. Below the ice the temperature to the bottom is relatively uniform and, as suggested by the temperature of 1.5°C at about 15 m, gives evidence of being rather thoroughly mixed-the mixing having occurred in the fall before the ice formed. The density is nearly uniform below the ice.

Because of the low angle of the winter sun and the shading of the water by the snow and ice cover, photosynthesis is inhibited. This, together with respiration of organisms and lack of oxygen replenishment from the atmosphere, results in low oxygen content of the water. Indeed, by late March oxygen is nearly depleted toward the bottom.

In the spring, a higher sun and increased day length brings about the melting of the ice. As the surface waters warm up to 4°C and become more dense, a slight, and temporary, stratification exists which sets up convection currents.

These currents, aided by wind, serve to mix lake water throughout until it is uniformly at 4°C and at its maximum density. Note from the vertical isotherms in Figure, that in Lake Mendota the mixing continued into early May even while the water was warming to 10°C. It is apparent that the amount of dissolved oxygen had increased throughout the lake. This vernal process of mixing has been termed the *spring overturn, or spring circulation period*.

Now we have seen that heat is absorbed very rapidly in the surface waters and, therefore, cannot be responsible for heating the deeper waters.

Thus some other agent must contribute to the circulation of warm waters into the deeper regions. *Heat transfer in lakes is accomplished mainly by winds*. It is only by this force that mixing can continue beyond 4°C.

Winds blowing across the surface of water set up a current resulting from frictional differences between moving air and water. Upon reaching the shore this current moves downward and across the bottom, thereby setting up the spring circulation.

As summer approaches, the weather warms, the longer days mean longer periods of insulation, and the brisk spring winds subside. Under these conditions the surface waters warm rapidly, expand, and become lighter than the lower waters.

Although the wind may continue to blow, its contribution to mixing of lake waters diminishes, for now the thermal density gradient opposes the energy of the wind. With the progression of the summer season the resistance to mixing between two layers of different density (resulting from increased temperatures) becomes greater than the force of winds—the significant density differences having been built up during periods of summer calm.

Reference to Figure reveals that Lake Mendota became thermally stratified during June and July. In late July, the uppermost region of warm, homothermal water extended downward some 8 m.

This region is termed the *epilimnion*. Below the epilimnion there developed a zone of rapid drop in temperature with depth.

This zone, only 2 to 3 m thick, is the *thermocline*. In modern limnological parlance the thermocline is defined as "the plane of maximum rate of decrease in temperature." The *zone* of rapid drop in temperature, including a gradient on either side of the thermocline, has recently been termed the metalimnion.

The condition of thermal stratification is very stable. The metalimnion constitutes an effective barrier between the epilimnion and the hypolimnion.

Currents stimulated by wind, as well as convection currents derived from cooling at the surface, are limited in depth by the metalimnion.

This means that heat and nutrients present in the region of maximum light are prevented from mixing throughout the body of the lake. Similarly, hypolimnetic water substances and many nonmotile organisms and those of limited mobility are restricted to the region.

For photosynthetic organisms this state of affairs is particularly critical. Depending upon the depth of the metalimnion, of course, photosynthesis is essentially inhibited in the hypolimnion.

This means that oxygen available for animal respiration is reduced, carbon dioxide is increased, and that microscopic algae (phytoplankton) as food for herbivorous animals is scarce. Indeed, it has been found in many lakes, Fayetteville Green Lake, New York, for example, that minute animals (zooplankton) eat great quantities of purple bacteria which are abundant in the deep, oxygenless waters.

The "impedance" to over-all lake current flow presented by the metalimnion has been called the *thermal resistance to mixing*. What, we might ask, is the character of this resistance, and how is such a barrier maintained? The answer to the first question is found in the density-temperature relationships mentioned briefly a few paragraphs back.

In Figure we see that between depths of about 15 m and 30 m the temperature declines from near 20°C to approximately 8°C, or about 12°C in 15 m. Recall that the density of water is dependent upon temperature, and that the difference in density per degree change is fairly great.

For example, as water warms from 4° to 5°C the density changes 8×10^{-5} gm mass/ml. Therefore within the rapid and wide temperature change in the metalimnion a considerable range of density also exists. Consider also the density difference between the temperature at the epilimnion-metalimnion interface and that at the hypolimnion-metalimnion interface in Figure. This difference represents a strong value of 0.00164 gm mass/ml. Where temperature-depth data are available a graph of the *relative thermal resistance* can be drawn. This parameter represents the ratio of the density difference between water at the upper and lower faces of each stratum to the difference between water at 5° and 4°C.

We might conclude then, that resistance to mixing presented by the metalimnian region is primarily the result of temperature decrease with depth and the related density differences within the zone.

Now, what perpetuates the metalimnion and its resistance to mixing? We have already learned that the epilimnion is a region of relatively free circulation and considerable turbulence, and further, that the metalimnion acts to limit the downward extent of water motion.

As wind-blown surface currents move against the shore of the lake they are deflected downward and encounter the metalimnion.

The metalimnion barrier in turn causes the current mass to spread out horizontally at the upper zone of the metalimnion. Each increment of warming of surface waters in the summer adds a proportionate measure of relative thermal resistance between the epilimnetic and metalimnetic waters. Thus the barrier effect is somewhat self-perpetuating in the sense that density differences between the two regions are increased as the continually warning waters circulate toward the cooler metalumnion.

Throughout the great range of lake morphology, climate, and topography, the various attributes of the thermoclune are subject to considerable variation.

The thickness of the metalumnion may fluetuate with season, becoming thinner as summer progresses. This is shown rather clearly in Figure, but in other lakes the effect may be more pronounced, the thickness ranging from about 12 m in early summer to around 5 m in late summer.

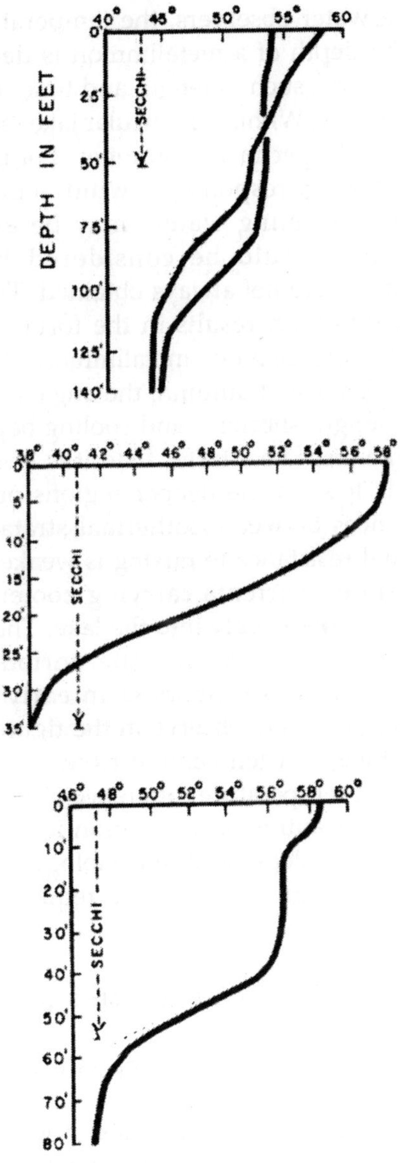

Fig. Temperature Profiles for Three Lakes of Various Depths

Recently a very thin metalimnion of some 2 to 5 mm in barber's Pond, Rhode Island, was reported. Across this layer, visible to underwater observers, the temperature decreased almost 8°C. The depth of a metalimnion is determined by a number of influences such as length and time of seasons, and by basin morphology. Within a particular lake the metalimnion usually becomes deeper in late summer. Local variations in metalimnion level in response to wind action and to the temperature of incoming waters may be expected. One additional point should be considered here; smooth temperature curves are not always obtained. Typically a late-developing stratification results in the formation of a thick epilimnion and more than one metalimnion.

With the approach of autumn, the angle of incident light decreases, day length shortens, and cooling begins. This is to say that the lake loses heat faster than it is absorbed. As the cooling extends toward the deeper regions of the lake, the density differences between isothermal strata become less. Thus the thermal resistance to mixing is weakened.

Windgenerated currents carrying cooler, oxygenated water reach ever more deeply into the lake. The metalimnion sinks rapidly. Overturn during the period of *autumnal circulation* began in Lake Mendota in early October and continued into December. Observe in the figure the uniform, and relatively high, oxygen content from surface to bottom accompanied by a temperature difference of less than 10. The period required for autumnal circulation is dependent upon local climate and the depth and morphology of the lake. The deeper the lake, the greater the time required for uniform cooling.

Two points are important here:

- Cooling (as with heating) takes place only at the surface,
- Vooling to 4°C is accomplished primarily by convection currents.

An *inverse temperature stratification,* not depicted in our figure, may exist during winter as the water cools below 4°C. Remembering that water reaches its most dense state at 4°C, we can easily appreciate that as surface waters cool beyond that point another "density difference" occurs in which the lighter, cooler waters float on more dense, "warmer" waters. Although the winter temperature conditions of stratification are reversed with respect to summer, the density relationships are similar—the less dense mass occupies the upper part of the lake.

The inverse stratification is most pronounced following the formation of the ice cover. Immediately below the ice (0°C) the temperature of the water rises sharply to near 4°C, or whatever the temperature of the water body. Inverse stratification is usually of rather short duration and, indeed, may not occur every year in a given lake. The period of *winter stagnation now exists.*

The sources and processes involved in the heating of lake waters beneath ice have received considerable attention. It appears, however, that heat from direct solar radiation through the ice, and heat derived from bottom muds are the major sources of energy for warming the lake. The heatdistributing mechanisms are apparently density currents of two types:

- The temperature-dependent currents which are derived from heating of water through the ice in the shallow zones and which flow into the deeper parts of the lake,
- Chemical density currents derived from dissolution of bottom substances in water; more dense by virtue of the dissolved materials, these currents may also transport heat.

THERMAL CLASSIFICATION OF LAKES

As a result of extensive studies of temperature phenomena, heat, and stratification, since the time of Forel, a great store of terminology and schemes of classification of lakes based on thermal characteristics has accumulated. In an attempt to keep terms and classifications at a minimum we have adopted a scheme proposed by Hutchinson.

By no means universally applicable, by virtue of the variable nature of lakes the system is useful and points to the more modern concepts and approaches to thermal properties and dynamics in lakes. Earlier terms and classifications, certainly not useless, are to be found in older texts and other works on ecology and limnology.

The following types are proposed for lakes occupying basins of sufficient depth to allow for stratification mixing ("mixes"), and the formation of a hypolimnion. Consideration has been given to altitude, geographical location with respect to latitude, and to depth of the basin.

Amictic

Lakes insulated and protected by ice cover from outside influences of weather and other factors; poorly known lakes of the Antarctic and high altitudes.

Cold Monomictic

Lakes of the polar regions in which the waters at any depth never exceed a temperature of 4°C; ice-covered and exhibiting an inverse temperature stratification in winter; one mixing ("monomixes") at temperatures not greater than 4°C in summer.

Dimictic

Lakes in which two circulations take place each year in spring and autumn; thermal stratification is inverse in winter and direct in summer; typical of lakes in the temperate zone and in higher altitudes in subtropical regions.

Warm Monomictic

Lakes of the warmer latitudes in which the temperature of the water never falls below 4°C at any depth; one circulation each year in winter, directly stratified during the summer. Lake Providence, Louisiana, stratifies from May through September, and circulates continuously from October to April.

Oligomictic

Warm lakes in which the water temperature is considerably higher than 4°C; circulation occurs rarely at irregular periods; found typically at low elevations in tropical zones.

Polymictic

Lakes in which mixing is continuous but at low temperatures, usually just over 4°C; characteristic of high mountains in the equatorial regions. Stratification does not develop because of heat loss to a relatively uniform environmental temperature.

HOLOMIXIS AND MEROMIXIS

With respect to circulation, or mixing, generally, most lakes are said to be holomictic ("wholly mixing"); that is, if circulation takes place it is complete and extends the entire depth of the lake.

In these lakes the temperature of the hypolimnion usually decreases uniformly to the bottom. However, some lakes are known in which the summer temperature curve resembles that of a holomictic lake except that temperature increases slightly through a considerable distance. How is it that warm water is lying underneath cold water? What of the density relationships? This bottom water is not less dense than the upper water.

It has been found that the bottom stratum usually contains large quantities of dissolved salts, and these increase the density sufficiently to prevent mixing with the upper layers. We have, then, a lake in which circulation is not complete; such lakes are called *meromictic* ("partly circulating"). The bottom, noncirculating layer is called the *monimolimnion*, and the density gradient becomes a *chemocline*. The waters above the chemocline in which thermal stratification such as in holomictic lakes can occur is termed the *mixolimnion*.

A lake in western Austria (Längsee) possesses a monimolimnion that has not mixed with the upper waters for over 2000 years. The monimolimnion contains a heavy concentration of salts derived from the sediments by biochemical processes. Heat in the bottom layer is apparently gained from bacterial activity and from insolation. Since there is no circulation by currents, heat is lost mainly through conduction. The annual temperature cycle of a meromictic lake is shown in.

HEAT BUDGETS AND LAKE STABILITY

Having considered the annual temperature cycle in a typical lake of the temperate zone, as well as some unusual cases of stratification, let us now give attention to some of the broader aspects of heat and heating. As stated earlier, temperature is the expression of the intensity phase of heat energy, while heat, as such, is an energy factor.

For purposes of comparison of like qualities and expression of the nature of the lake waters as environment with respect to organisms, the calculation of heat income and heat budgets becomes a useful tool. *Summer heat income* is the quantity of heat delivered to a lake necessary to warm the waters from the homothermal spring condition of 4°C to the summer maximum. Although direct solar radiation contributes essentially all of this heat, the distribution of it is accomplished by winddriven mixing of waters of differing densities. The work of the wind in summer heat income is done against gravity because of density differences arising from heating of surface layers by direct insolation.

The *annual beat budget* takes into account the total quantity of heat taken into the lake to warm the waters from the lowest temperature of winter to the maximum summer temperature. Either parameter can be determined in absolute terms such as calories, or gram-calories. Usually, however we wish to compare heat characteristics of several lakes, in which case it is well to introduce area. The result is then expressed in calories per square centimeter of water surface. The heat content per unit surface area of a lake at any given time can be determined by totaling the product of temperature times volume for declared depth intervals, and then dividing by the surface area of the lake. The gain in heat content per unit area beyond the hemothermous 4°C state is the summer heat income, and is more a function of mean depth than of temperature differences. The annual heat budget may be derived in the same way, except that the lowest winter temperature is substitute for 4°C. Table gives morphological and thermal data for five lakes in the Convict Basin of California, and shows the relationship between depth and annual heat budget.

Comparisons of annual heat budgets of lakes must consider depth differences. Shallow lakes are not able to store all the heat transmitted into them. During the period of maximum temperature, the lakes become vertically homothermous and heat is transferred and lost to bottom sediments.

Within temperate regions the annual heat budgets of typical lakes range from about 30,000 to 40,000 cal/cm^2. Much of this heat (33 to 75 per cent) enters the lake after the spring circulation period, and work energy is involved in the circulation of the heat to lower lake levels. lake Michigan, with an annual heat budget of 52,400 cal/cm^2, and Lake Baikal, whose annual budget amounts to 65,500 cal/cm^2 are the highest known. The summer heat income for Lake Michigan is 40,800 cal/cm2, and Lake Baikal receives 42,300 cal/cm^2 between spring isothermal 4°C and summer maximum. To bring about the thermal condition shown in Figure for Lake Mendota an annual budget of about 23,500 cal/cm^2 and a summer income of some 18,240 cal/cm^2 are required.

Table. Major Physical Features of Certain Lakes

					Temperature, °C	***Heat Intake, gm cal/cm^2***		
Lake	**Elevation**	**Surface Area (ha)**	**Mean Depth (m)**	**Transparency (m)**	**Bottom to Surface**	**Lake Mean**	**Summer Heat Income**	**Annual Heat Budget**
Convict	2275	68.6	26.4	15.3	6.6-14.7	10.9	18,173	22,984
Mildred	2970	4.3	8.1	12.0	10.0-12.5	11.1	5,822	7,298
Witsanapah	3228	1.8	8.4	9.0	5.5-10.0	8.5	3,827	5,358
Edith	3030	7.3	17.7	4.1-14.1	13.8	8.7	8,601	11,826
Dorothy	3102	43.8	40.8	20.1	4.4-14.7	8.8	17,967	25,402

In view of the importance assigned to the work of wind in distributing heat in a lake , we should question how much work a wind of a given velocity can do and how much work is required to mix a given quantity of heat. As Hutchinson has pointed out, little work has been done on the problem, and our knowledge is therefore meager.

However, some data from various sources as compiled by Hutchinson will serve to convey a general idea. With respect to the first question , Langmuir has stated that winds of velocity 300 to 700 cm/sec developed forces of 0.65 to 6.3 dynes/ cm^2 on lake surfaces. The summer heat income of Green Lake, Wisconsin, is 27,316 cal/cm^2 and the heating period is 122 days. Calculations show that the mean rate of work directed toward heating the waters is 0.02 dynes/cm^2. These data indiçate that the work of the wind in summer heat distribution is very small in comparison to the force of wind on the lake surface.

The loss of heat during the period of cooling from the summer maximum temperature to the winter minimum must be of the order of the annual heat budget. This loss from the lake is mainly through radiation to the surroundings. Cooling by evaporation also expends heat. The waters of effluent streams remove considerable quantities of heat from lakes, especially from impoundments where warm, surface waters are removed while cooler waters remain behind.

From the foregoing considerations we sense that a certain quantity of energy is expended in developing summer stratification in holomictic lakes. We have also made reference to the stability of the stratification and to the resistance offered by the lake to mixing. This, in essence, is one definition proposed for the term *stability*.

More properly, *stability of stratification* is that energy of resistance which the lake offers to oppose upset of density stratification. As warm, less dense water comes to overlay cool, more dense water during stratification the centre of gravity is shifted downward. Stability then becomes a measure of the amount of work required to raise the centre of gravity or to displace it to its original position. Stability decreases in autumn as the thermocline sinks below the centre of gravity and the waters above and below this centre attain similar densities. The greatest stability is probably reached just prior to maximum heat content in summer.

While we have been considering temperature and heat relationships in relatively large bodies of standing waters, we should not lose sight of the fact that pools and small ponds also exhibit responses to seasonal and daily temperature influences. Such responses often limit activity of some organisms and should certainly be studied in limnological investigations of small situations. "Microthermoclines" and odd temperature stratifications exist in certain small bodies under given conditions. A reversed stratification has been shown to exist at times in small Burt Pond, Michigan.

Table. Summer Pond-water Temperatures

	°C
At surface of Mat	21 to 22
2.5 cm Below Mat Surface	23 to 24
5.0 cm Beneath Mat	25
7 to 8 cm Beneath Mat	21 to 23
15 cm Beneath Mat	20

The data for this pond are given in Table and indicate, among other things, how rapidly the temperature of a small body of water responds to atmospheric conditions. It is suggested that the reversed stratification results from heating of the surface by direct radiation in the morning, followed by cloudiness and winds which cool the surface in the afternoon.

THERMAL PROPERTIES OF STREAMS

The basic thermal characteristics and processes in the water in streams are not different from those of lake waters. Penetration of light, absorption, specific heat, and other factors attributable to the nature of the water molecule operate in determining the fitness of the stream as an environment just as in lakes.

As an environment, however, a stream presents a considerably different set of temperature conditions. These conditions derive from variations in velocity, volume, depth, substrate, cover, water source, and a number of additional features operating seasonally, daily, and even longitudinally along the stream course at a given time. Recall that we have termed a stream community an "open system" by virtue of the considerable land-water interrelationships; some of these influence temperatures.

The major factor in the warming of stream waters is direct solar radiation. A series of temperature curves recorded over several successive days of cloudless skies followed by an overcast day will show rather regular fluctuations from late afternoon maxima to early morning minima during the cloudless period, but the cloudy-day curve will be flattened. Additional evidence for the importance of direct insolation is found in the frequent occasions when the temperature of the water exceeds that of the air.

This usually occurs on very clear days of intense sunlight. The limit of nocturnal cooling is, of course, regulated in small, shallow streams by atmospheric temperatures. In larger, deeper streams it depends upon the rate of heat loss before warming begins again. Additional heat may be gained from bottom sediments in streams, but little data are available to show the extent of such heating.

On the other hand, the temperature of stream water is a measure of the actions and interactions of a wide variety of factors. Consider a stream rising in a rocky, wooded highland and flowing down over the coastal plain to the sea. In the upper reaches the waters are cooled by the substrate, by the shading provided by vegetation, and possibly by the entrance of spring-fed tributaries.

Turbidity may be quite low. As the stream approaches the lowlands it becomes wider and deeper, and more water is exposed to direct sunlight. This, and increased silt content which absorbs considerable heat, result in the development of a segment quite different with respect to temperature from the upper reaches.

Depending upon size and origin of streams, their diurnal and seasonal temperatures follow atmospheric temperatures more closely than do those of lakes. Diurnal temperature variations at a given point in a stream may be related to two major sets of factors:

1. Conditions at that point,
2. Conditions upstream from the point.

Under (1) we should consider velocity and discharge, season and hour, and the daily range of fluctuations of air temperatures at the point. Factors pertaining to (2) include the nature of upstream environment, substrate (and impoundments, if any), atmospheric conditions and temperatures upstream, and distance and time of flow from critical upstream situations.

With respect to size and water temperature fluctuations, we can conclude that the smaller the stream, the greater the temperature variations and the more rapid the response to environmental fluctuations. Furthermore, these fluctuations exist throughout the year, but are minimized beneath ice cover.

The range of daily variation of water temperature is maximal when there is the greatest differential between mean diurnal air temperature and mean water temperature. With increased volume and turbidity the range of fluctuation diminishes. Because of the wide range of temperatures exhibited by small, cold streams, a single temperature reading may not at all approximate the average daily figure.

The source of the stream's waters and the nature of the drainage pattern in some instances determine the thermal properties of the stream. Many spring-fed streams and those originating from surface discharge of subsurface aquifers are essentially thermostatic. Lander Springbrook, New Mexico, is a small rheocrene ("flowing") spring. Seventy meters below the source the annual variation in water temperature is 4°C; air temperatures fluctuate from about 14.3° to 33°C.

The annual variation of water temperature of Silver Springs, Florida, a large spring run (daily discharge about 600 million gallons mainly from aquifer openings) varies no more than 1° throughout the year. Diurnal fluctuation in Silver Springs is also of the order of 1°. These spring-fed streams do not receive significant quantities of surface water which might otherwise bear on their thermestatic qualities. Surface-fed streams typically show wide seasonal fluctuations corresponding with atmospheric conditions.

Because of turbulence and the shallow nature of most streams, thermal stratification is not generally an attribute of streams. When stream waters stratify, the process usually takes place in pools along the stream course. In some instances the stratification is a result of inflow of cool spring water at the lower levels of the pool.

In Jack's Defeat Creek, Indiana, a pool was found to be stratified, the temperature decreasing 7.5°C from surface to bottom at about 62.5 cm. Other qualities, including free carbon dioxide, dissolved oxygen, and alkalinity were also stratified. Inflow of spring water possibly accounts for the local stratification in this case. If warmed upstream waters are gently introduced into a cool pool of water, the inflowing, less dense waters are apt to flow over the more dense pool waters and on downstream without considerable mixing.

One word on the effects of human endeavors, such as agriculture and industry, on stream temperature and ecology needs to be interjected here. We know that water taken from a stream and used in irrigation is warmed in the process and, upon being dumped back into the stream, increases the downstream temperature for some distance.

Similarly, water used in cooling various industrial installations warms the stream. In addition to the simple warming of the stream which, as such, may exceed the temperature tolerance of some of the organisms, the increased heating also decreases the oxygen retention capacity of the water, thereby affecting certain organisms and stream metabolism. Reservoirs, whether for water supply or hydroelectric power, exert considerable influence on the quality of stream water below the impoundment.

As spring temperatures of streams approach the favorable point for spawning of warmwater fishes, sudden discharge from the lower levels of an upstream impoundment cools the stream and inhibits reproduction of the fishes. Conversely, the influx of warm, surface waters from an impoundment into a cool trout stream has deleterious effects on the fishes and the stream community. Much precise work needs to be done on these aspects of stream relationships. The results of such research could go far in enlightening the agencies concerned with dam-building projects.

HEAT AND TEMPERATURE IN ESTUARINE WATERS

Before considering some of the more general thermal aspects of the estuary as a body of water, we should give attention to the effect of increased concentration of salts upon certain basic properties of water. Recall that pure water has a specific heat of 1.

As the concentration of dissolved salts increases, the specific heat decreases. For sea water at 17.5°C and salinity, of 35‰ (parts per thousand) the specific heat is 0.932. This suggests that less heat is required to warm a given volume of salt water than to warm the same amount of fresh water. Whereas for pure water density is a function of temperature alone, being maximum at 4°C, the presence of salts depresses the temperature of maximum density in sea water.

At a salinity of 24.70‰ maximum density occurs at the freezing point, -1.33°C. At salinities higher than 24.70‰ the temperature of maximum density is below the freezing point. Unlike pure water, sea water at high salinities undergoes regular increase in density as it cools to freezing.

Generally, the freezing point of sea water varies inversely as the salinity. At a salinity of 10‰, commonly experienced in estuaries, the freezing point is near -0.5°C; at 35‰ sea water freezes at about -2.0°C. The complexities in estuarine thermal dynamics are easily appreciated when we remember that within a few miles the salinity in an estuary may grade from near 0‰ to 25‰ or higher.

The heat content of estuarine waters is derived mainly from solar radiation. These waters are directly heated *in situ* as they occupy the estuary basin. Heat is also received indirectly from inflowing stream water and from tidal flow from the sea. The temperature of the estuary is, therefore, primarily a function of the temperatures of entering streams and the sea together with tidal stages. If the estuary empties into a relatively deep sea, the seaward temperature is apt to be more stable than in the upper zone where the stream temperature may fluctuate widely. On the other hand, estuaries of spring-fed streams flowing into shallow sea zones may exhibit more stable features in the headward regions.

In view of the shallow nature of most estuaries we should expect considerable diurnal and seasonal fluctuations in surface temperatures. In the East Bay, an arm of the Galveston Bay system of Texas, during July, surface water warms from an early morning temperature of about 27.0°C, to 33.0°C in late afternoon; atmospheric temperatures during the same period range from near 28.0° to 34.0°C. Seasonal temperature fluctuations depend upon latitude and a number of local factors such as water source, basin morphometry, winds, and tides.

In Apalachicola Bay, Florida, the annual temperature range is of the order of 25°. Temperatures in the Sheepscot estuary of Maine range from about freezing to near 25°C in the upper reaches of the estuary, while in the lower region the range is about 15°. These temperatures are determined primarily by atmospheric and climatological conditions. A number of instances have been noted, however, in which sudden changes in the pattern of oceanic currents result in the influx of cold masses of water into an estuary, thereby drastically reducing water temperature.

Another factor that affects temperatures in estuaries is the flooding of marshes and mud flats during high tides. In warm, humid regions the surface of exposed marshes may, during the summer, become quite warm. As the rising tide covers the marshes with estuarine waters, heat is imparted to the waters.

In other regions the exposed muds may become cooled through evaporation and thus lower the temperature of incoming water. Studies in the Elkhorn Slough estuary of California have shown that mud gains heat much less rapidly than the over-lying water, even in a depth of only 15 cm.

Temperature distribution in estuaries is largely a function of depth together with the relative effects of stream inflow and tidal exchange. In a shallow, mixing estuary the waters tend toward vertical homothermy. Longitudinal temperature patterns vary seasonally.

Depending upon the morphology (surface-volume ratio) of the stream basin and the volume and rate of discharge, the upper waters of the estuary may be cooler in winter and warmer in summer than the lower estuary waters.

From the mouth of the Patuxent River of Chesapeake Bay to a point about 74 km upstream, the average surface temperature from March to July grades from about 23° to 21°C; bottom-water temperatures similarly decrease on the order of approximately 3°. From September to January, both surface and bottom temperatures increase slightly over 1° from the bay upstream.

Where conditions of climate and sufficient depth are met, estuaries may exhibit a vertical temperature gradient. In summer the surface waters are usually warmer than the underlying layers. With the cooling weather of autumn, the upper waters cool more rapidly than the deeper masses, resulting in overturn and mixing. In midwinter the temperature of the surface waters may hover near that of maximum density, thereby giving rise to convection currents which continue to circulate the waters.

Increasing insolation and winds in spring warm the surface, and a temperature gradient follows. Density stratification derived from cold or warm fresh water flowing over the more dense salt water also contributes to vertical temperature differences in estuaries. Such a stratification may be recognized by salinity measurements from the surface downward.

Chapter 3

Natural Waters in Motion

In the smallest pool and the largest lake or stream, movement of some form and degree takes place. Motion in such waters derives from temperature and density differences, gravity, and wind acting together or singly and to varying extent.

Directly or indirectly every facet in the aquatic community, be it physical, chemical, or biotic, is in some way affected by movement within the body of water.

In this chapter we wish to single out some of the more conspicuous forms of motion in lakes, streams, and estuaries and to consider them as basic features of natural waters and as physical factors determining the fitness of waters as an environment for habitation by living organisms.

MOVEMENT OF WATER IN LAKES

As a general rule, the water of a lake basin is partially or wholly in motion. It is this movement, derived from either internal or external forces, or both, that is responsible for the circulation of heat, dissolved substances, and some organisms in lakes.

Turbulent flow is characteristic of the movements of lake water; this results from the action of molecular systems imparting irregular direction and velocity to water mass.

Turbulent movement is then incorporated into the more conspicuous forms of motion in lakes. The larger movements of water are called *current systems*, and are frequently defined as two types: nonperiodic, or arhythmic; and periodic, or rhythmic.

NONPERIODIC CURRENT SYSTEMS

Nonperiodic systems are those often termed simply, "currents," implying a unidirectional flow of water. This movement may be caused by differential heat distribution within the lake, the passage of stream waters through the lake, and by winds. Broadly speaking, nonperiodic systems are produced and maintained by external forces.

Wind is doubtless the major external force acting to set lake waters in motion. Sustained winds blowing across the surface serve to pile up water at the down-wind end of the basin. The result is lowering of the water mass in another part of the lake. As the winds subside, the piled-up water begins to flow along a gradient, or down the slope, so to speak. Eventually the over-all lake level reaches equilibrium and the current system comes to rest. We have seen that the action of wind is also important in the development of summer stratification of lakes.

A phenomenon familiar to all who have traveled the ocean or large lakes is that of *wind streaks*, or those linear accumulations of foam, floating debris, or oil. Irving Langmuir, better known for his work in electricity and related fields, demonstrated that surface flow resulting from wind drift is not an even and uniform movement of water containing variable velocities, but rather possesses a definite pattern.

Water flow in wind drift is in the form of helices lying parallel to one another oriented in the direction of the wind with the direction of the helices alternating clockwise and counterclockwise, and the effect of streaks is derived from the accumulation of materials in zones of convergencies between clockwise and counterclockwise helices. The width of the zone of convergence varies seasonally, being wider in the late autumn than in spring. It appears that the downward vectors of the helices serve to transfer momentum from the surface into the deeper regions of the lake.

In large lakes the effects from stream inflow, wind, and earth's rotational forces often combine to produce a pattern of large current swirls. The surface currents of Lake Constance, in Europe, have been studied rather intensively, especially with relation to the path of the Rhine River through the lake. As a result of geostrophic forces and the shape of the basin, the Rhine waters flow across the upper end of the lake and follow the northern shore. The river current, together with wind patterns, apparently produce the large swirl between Langenargen and Rorschach.

The influence of the Rhine on the currents in Lake Constance serve to point up another kind of nonperiodic current in lakes; such may be termed *density currents.* These currents result from the passage of river water of a given density through a lake of differing density.

As suggested previously, the route of the stream water is determined mainly by geostrophic forces, the shape of the lake basin, and local meteorological agents. The depth at which the river water flows is a function of density differences.

If the density of the river water is greater than any of the lake waters, the stream flows over the bottom of the lake. Similarly, if the lake is generally more dense than the stream, the latter flows at the lake surface. During the summer months the water of the influent is usually cooler than the lake surface, and therefore more dense.

Thus the river water flows downward until it meets a region of greater density, whereupon the river flow becomes horizontal. The "waterfall" at the mouth of the Rhone in Lake Geneva is familiar. In thermally stratified lakes, the river flow is usually above the hypolimnion, thus the lower lake levels receive little of the value of enrichment by the inflowing stream.

In Norris Lake, a Tennessee Valley Authority impoundment, there are indications that density currents move through the lake as the water is drawn down for hydroelectric power.

These currents may be poor in dissolved oxygen content due to organic decomposition, causing movement of fishes to avoid the oxygen minima. Density currents may be important in ponds and small lakes, and probably relate to distribution of materials beneath ice cover in winter. In Lake Mead, formed by Hoover Dam on the Colorado River of southwestern United

States, a density current resulting from increased silt load forms during the summer. This *turbidity current* may flow into the lake below the level of the river density current.

One more aspect of nonperiodic movement of water should be considered, that of *turbulence* and *eddy effects*. Although the mechanics of these phenomena are best known from laboratory studies, the applicability of the principles to limnology is vast and deserving of more intensive study in the future. To anyone who has watched a stream plunging through riffles or along irregular shores, the intense turbulence and the resultant eddy systems are obvious. Within lakes, however, such turbulence is reduced and certainly less conspicuous, albeit present as the result of currents such as those previously described and to be discussed later. The great importance of eddy effects is in heat transfer within the water mass (*eddy conductivity*), in the diffusion of dissolved materials (*eddy diffusivity*), and momentum transfer (*eddy viscosity*).

We distinguished laminar and turbulent flow with respect to streams. It is agreed that laminar flow, wherein sheets of liquid move uniformly without local variations in velocity (but with velocity differences from "sheet" to "sheet"), does not normally occur in nature. The more characteristic flow of natural waters is one of turbulence, in which irregular motion of subsidiary masses is contained within some relatively simple current. The nature of the turbulence is an expression of various factors including flow velocity, velocity gradients encountered, and the nature of the boundaries of the flowing systems.

Figure may be interpreted in terms of a density current passing through a lake. In this example, the lake proper serves in somewhat the same sense as a stream bank in offering resistance to moving masses. The velocity and magnitude of the eddy systems operating at the upper and lower zones of the density current, for example, are functions of velocity factors of the lake proper and the current, and boundary characteristics including temperature and chemical composition. It appears, then, that through eddy effects, mass and molecules can be exchanged from one layer to another.

One effect of eddies is that by which masses leaving one layer carry with them momentum from that layer and acquire the momentum of the new layer before returning to the original one. This is a simplified definition of *eddy viscosity*. In practice, a *coefficient of eddy viscosity* may be calculated from:

$$r_s = -A\frac{d\bar{u}}{dn}$$

in which rs is the Reynolds stress, or shearing stress per unit surface area;

$$\frac{d\bar{u}}{dn}$$

is the shear of the velocities as measured; and A (from "Austausch") now becomes the coefficient of eddy viscosity, that is, the relationship between mass and transverse distance through which a liquid flows per unit time in a path corresponding to that of the main direction of flow. The transfer of motion through eddy viscosity increases with turbulence, thereby giving rise to the fact that eddy viscosity is considerably greater than molecular viscosity, often on the order of 1000 to 1 million times. In natural waters eddy viscosity is a primary force in reducing the settling rate of sediments, and in maintaining the position of microscopic organisms.

In the same sense that eddy viscosity relates the transfer of mass and momentum from one water stratum to another, *eddy conductivity* describes the exchange of heat across surfaces. This parameter takes into account the specific heat of the fluid and the heat gradients encountered between masses. This heat transfer is then proportional to the mass transfer described in the formula for eddy viscosity. Important here is the idea that conduction may be lateral as well as vertical. Eddy conductivity is fundamental in the heating processes of lakes, described in the preceding chapter.

The transfer of dissolved substances, so very important in lake metabolism and maintenance of living organisms, is another function of mass exchange and is termed *eddy diffusivity*. An important concept of eddy effects is that of the transfer of more than one substance, or the *principle of common transport*. This principle holds that a motion that transfers one environmental factor may, simultaneously, carry another property.

For example, while dissolved salts are being transported downward, dissolved gases, or heat, may be carried upward in the same eddy system. Some idea of the rate at which substances may be distributed through this process is gained from the fact that summertime eddy diffusivity in the western part of Lake Erie is near 25 cm^2/sec, or probably some 104 times that in water in a quiet bottle.

The basic concern with eddy effects for our purposes relates to the more general ways in which the processes contribute to lake dynamics.

PERIODIC CURRENT SYSTEMS

To this category belong water movements exhibiting some form of rhythm or periodicity. Two major systems are to be considered; these are: traveling, *surface waves* and standing waves, or *seiches*. As has been shown for nonperiodic currents in the preceding section, external forces, mainly wind and other meteorological factors (atmospheric pressure), are responsible for periodic currents.

SURFACE WAVES

Lake surfaces are never completely smooth; slight irregularities and turbulences are present. Friction occurs between these surfaces, and wind blowing across the lake results in movement of the water. This motion imparts the first phase of wave formation. Subsequently, a crest is built, and increased wind velocity tends to build up the wave; eventually, an eddy is produced by wind on the trailing slope of the wave, giving a certain thrust to its back, thereby accelerating the motion of particles in their orbits within the wave.

There is no essential horizontal movement to waves in open water. A particle of water moves in a vertical circular pattern, or orbit, returning to its original position as the wave rises and falls. The diameter of the orbits and, correspondingly, the velocities, decrease rapidly with depth. It has been calculated that the diameter of orbits at a depth equal to onehalf the wave length is on the order of 1/500 that of surface orbits, or quite imperceptible. Not until the wind develops to such a force as to produce a tumbling crest, or whitecap, is there any forward motion of the water. A similar effect is exerted by the bottom in shallow zones where friction directed upon a wave causes the mass to pitch forward, producing *surf*. These breaking waves are called *waves of translation*; open-water swells are termed *waves of oscillation*.

Both velocity of wind and the time that it has been exerting force upon a wave determine the size of waves. In order to build waves of any great amplitude a considerable amount of time is required to transfer the wind energy into the kinetic and potential energy of a wave. Similarly, space is required, for a wave must have area in which to move before the wind in order to gain amplitude. The relationship between wave height and fetch during strong winds is expressed in:

$$b_w = 0.105\sqrt{x}$$

in which x is the fetch, or downwind distance (in centimeters) from shore to location of the wave in question.

Although waves and wave action are of especial interest to physical limnologists, our interest in terms of the over-all ecology of lakes is mainly in the action of waves in circulation of vital materials in lake waters, and the stirring effects of breaking waves in shallow-water zones. From a broader point of view, wave action is most instrumental in sedimentation and erosional processes which serve to modify the morphology of the lake basin and littoral areas.

SEICHES

A seiche is a form of periodic current system, described as a standing wave, in which some stratum of the water in a basin oscillates about one or more nodes. This rocking motion can be seen in a bowl of thin soup passed by a shaky waiter, the seiche being set up by movement of the basin. By blowing upon the soup near one side of the bowl one can create a seiche due to pressure changes, although the basin be fixed. Seiches in lakes have been recognized for many years, and since Forel's pioneering work, beginning about 1870, much attention has been given to this intriguing phenomenon.

The name, attributed to de Duillier, writing in 1730, is derived from the Latin *siccus* through the French meaning dry, and referring to the exposure of low shorelines left dry as the lake water recedes during an oscillation.

We have seen previously that sustained blowing of wind across the surface of a lake will cause water to pile up in the down-wind region of the basin. If the wind suddenly subsides a current will naturally flow toward the area of lowered surface. The water mass does not come immediately to equilibrium.

As a result of the energy of motion imparted by the gradient current, an oscillation occurs about a stationary node. This oscillation, a *surface seiche*, continues until damped by contact with the basin proper or by meteorological forces.

The origin of surface seiches may be found in a number of natural mechanisms. From the considerable study of lake oscillations, however, it appears that surface depressions resulting from sustained winds, and from local sudden changes in atmospheric pressure, are the primary agents. There are cases in which local rain showers set up a seiche due to the pressure of the impact of the falling rain. Sudden inflow from a contributing stream has been suggested as the origin of a seiche in Loch Earn, Scotland. Earthquakes have been named also as possible causes of seiches.

Since these standing waves operate according to the laws of oscillating systems, the formation of harmonics is a general characteristic. In addition to the simple, uninodal seiche, binodal and trinodal seiches have been reported commonly. In fact there is at least one record in Loch Earn of a seiche exhibiting a nodality of the sixteenth. Normally, the periods and locations of the nodes are determined by morphological features of the basin, i.e., depth, diameter, and form. The amplitude of the seiche depends upon the source and the intensity of energy giving rise to the oscillation and to the form of the basin.

Although seiches occur in all enclosed or partially enclosed bodies of water, in small basins the effect is detectable only by sensitive recording devices. In larger lakes the oscillations may be quite conspicuous.

Because of the narrow form of Lake Geneva near the city of Geneva, waterlevel fluctuations due to seiches often leave the shoreline dry. Here, in 1841, a seiche with an amplitude of 1.87 ni was recorded. In the same lake in 1891, Forel observed a seiche which lasted for 7 days, 17 hr. This standing wave underwent 150 oscillations, the amplitude dropping from a maximum of 20 to 7 cm; the period of oscillations was of the order of 73 min. The Great Lakes exhibit oscillations, the periods of longitudinal seiches in Lake Eric are near 790 rain, and for Lake Huron 289 min. Transverse sciches are also known and serve to complicate measurement of lake activities.

Amplitude and period of seiches are determined in practice by the use of a float contained in a protective housing. The float is connected to a counterbalanced pointer over an appropriate scale. Termed a *limnometer,* the device may employ a tracing stylus to give a graphic recording or limnograph. The periods of seiches can be calculated from various, and often complicated, formulas. For a uninodal seiche in a rectangular basin of regular bottom the period t may be calculated from:

$$t = 2L\sqrt{gh}$$

in which L is the length of the basin, b is the water depth, and g is gravity acceleration. Bearing in mind that a seiche is water in motion, it becomes apparent that turbulence will enter into the system as soon as oscillation begins. The effect of turbulence will be to damp out the oscillation as well as to increase the period.

Thus far, our discussion has dealt with the *surface seiche.* Since about the turn of the century, however, it has been known that internal, periodic current systems could be present in lakes under certain conditions. Within a stratified lake of two or more layers of differing densities the strata may oscillate with respect to each other without being apparent at the lake surface. Recent compilations of data relating to *internal seiches* indicate that both the ranges and the periods of such seiches are significantly greater than those of surface seiches. During summer stratification the thermocline may, as a matter of character, oscillate between the lighter layer of the epilimnion and the denser hypolimnion. An oscillation of this nature can be clearly and relatively easily detected through temperature recordings at various depths.

Data derived from laboratory and field studies of displacement of lake layers of equal temperature (isotherms) have led to the general acceptance of two major forms of internal seiches.

One type, the thermocline seiche, results from the movement of water masses between the epilimnion and the most dense zone of the metalimnion. Another system, the *hypolinmion seiche,* moves along the boundary between the metalimnion and the hypolimnion.

The time of one complete oscillation (period) is usually greater in the case of the hypolimnion seiche than for the thermocline seiche, both, however, being more or less dependent on the length of the lake. The differences in periods are due to unequal densities of the two strata. In Lake Geneva (about 37 km in length), the period of an internal seiche is about four days.

The mechanisms responsible for setting up interval waves are the same as for the surface seiche. Wind-driven waters of the lake surface pile up at the windward end of the basin, causing the deeper waters at the leeward end to compensate by being pushed up.

If the lake is stratified, however, the resultant effects of the displacement are different. Because of greater density difference between air and water than between cool and warm water, slight depression of the water surface by air in motion results in a proportionately greater displacement of the lower water stratum toward the surface in the windward end of the lake. As the wind diminishes, the isothermal strata oscillate about a node near the centre of the lake. The metalimnion serves as a sort of "cushion layer" on which the epilimnion moves over the hvpolimnion.

The problem of internal sciches is not a simple one, either in terms of seiche motions and dynamics or definition. All internal oscillations in lakes may not be "true" seiches, that is, motion resulting from free oscillations. Forced oscillations, or those that are sustained by direct wind action, may be present and sometimes mixed with true seiches. In very shallow lakes forced oscillations may damp the effects of seiches, making difficult a distinction between the two.

Internal seiches are of considerable importance in the total economy of the lake, probably more so than surface sciches. In Lake Victoria an internal seiche with a period of about 30 days brings about the displacement of water into remote parts of the lake basin. The resultant turbulence mixes lake water over the bottom materials and transports dissolved and suspended materials into the shallow zones.

In addition to this more or less horizontal transport, internal seiches serve to distribute heat and nutrients vertically within the lake. One particularly interesting problem is created by the action of internal seiches; this is the determination of the position of the thermocline through temperature measurements. The variations in position of the thermocline when in oscillation. It has become increasingly apparent that the thermocline is not a stable stratum, but rather is subject to much shifting and modification during the season.

True, or lunar tides, have been reported for large lakes, although the amplitude of lake tides would naturally be small. In Lake Superior the range is about 1 in. (3 cm). The difficulty in observing and measuring lake tides arises by seiche complication; a seiche in phase with lunar time (12.5 hr) would magnify the tide range, or if out of phase, would inhibit tidal action.

STREAM CURRENT SYSTEMS

The major processes and properties of flowing waters have been described the importance of such processes in the formation, maintenance, and alteration of the stream or channel. It is suggested that so much of that chapter as pertains to the dynamics of stream flow be reviewed at this time, for those same processes and attributes contribute to the several generalities to be considered in this section.

We have seen that wind is the major force in creating current systems in lakes. In streams the important factor in producing water movement is gravity. So long as a gradient on the earth's surface exists, water flows in response to gravity, to "seek its lowest level" following the route of least resistance.

Because the route is seldom straight, the gradient rarely uniform, and the sides and bottom of the channel not often smooth for any distance, varying velocities and turbulences result. Bearing in mind the importance of these factors and activities in channel structure, let us now consider them with reference to the stream as a living-space.

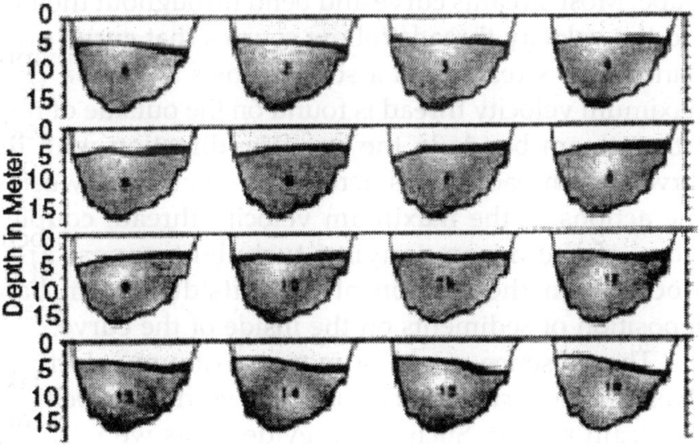

Fig. An Internal Seiche.

A conspicuous characteristic of stream flow is transport of dissolved substances, suspended materials, and living plants and animals. This ability to move matter also contributes greatly to the unstable nature of bottom types and shore zones, mainly through shifting of bottom materials and through sedimentation. Although these processes and features could result from nonturbulent flow, they are usually of greater magnitude in turbulent waters.

Turbulence in streams is not uniform throughout a cross-sectional area of the stream. Maximum turbulence occurs at some region in the cross section and diminishes from that zone toward the surface or the bottom. Zones of maximum turbulence are usually located on either side of a similarly restricted *thread of maximum velocity*.

The number of high velocity threads and turbulent zones apparently increases with channel width, depth, and nature of the bottom. The result is decrease in stream stability, making for a more rigorous environment for plants and animals. Eventually, however, uniform discharge may contribute toward a balance between the bottom materials and turbulence and the stream bed becomes more tenable for organisms.

Maximum velocity threads and associated turbulent zones do not maintain a constant position in the stream cross section, except perhaps in straight streams occupying uniform channel shape. Most streams curve and bend throughout their courses, and the velocity threads follow courses that curve even more than do the streams. As a stream flows around a curve, the maximum velocity thread is found on the outside of the curve; if the stream bends in the opposite direction with the next curve, the thread moves across the channel between bends. The actions of the maximum velocity thread, coupled with those of the accompanying turbulent zones, are major processes in the erosion of the outside stream bank and deposition of sediments on the inside of the curve.

The presence of one or more maximum velocity threads gives rise to varying horizontal and vertical velocities within the steam proper. Such are in evidence as we view the body of a stream and observe the morphology of the stream bed. We are impressed by the presence of a large sand bar separated by a deep channel from, perhaps, a gravel bed; farther downstream the bottom may be strewn with large rocks or be quite muddy.

These features attest to the ever-changing nature of the stream resulting from shifting of the velocity threads and the turbulent zones, and further, to the carrying capacity of the current itself as a function of velocity. Obviously, the relationships are not nearly so precise as indicated here. The velocity varies with the depth and with obstructions such as boulders. In other words, an organism attached to the bottom or to the downstream side of a large stone may not actually be inhabiting a current velocity indicated by the general nature of the stream bed.

Swirls and eddies of widely ranging proportions constitute another form of current systems in streams. These systems serve to mix stream waters and to effect deposition of organisms, organic debris, and sediments. Many forms of algae are adapted to eddy dispersal of vegetative structures by taking advantage of shallow, well-lighted swirls for rapid reproduction and subsequent distribution as the eddy contributes to the main current.

Density currents due to differences in temperature, chemical composition, or to silt load occur in streams, especially large streams with smooth bottoms. Spring-fed tributaries are often cooler than the main stream, and as the colder waters meet with the main stream a stratification results. In the lower reaches of a stream or, more properly, in the estuary, salinity currents move with the tides underneath the fresh water. Near the mouth of the Escambia River, in Pensacola Bay, Florida, salinity measurements made in the autumn indicated a surface salinity of 4.5 parts per thousand while at the bottom (about 4.5 m) the salinity was 24.4 ‰. This sharp salinity stratification was also reflected in the freshwater and marine fish distribution.

Tides exert influence on river flow, often well beyond the brackishwater zone. This effect serves to increase turbulence in the lower stream course, and also to bring about slight periodic flooding of the valley. This latter process increases the exchange between stream and land. The Paumunkey River flows on the Piedmont plateau and coastal plain of Virginia, and salinity effects are felt some 12.8 km from the coast. Tidal influences, however, reach about 64 km upstream.

From the consideration of fluvial processes, and the brief review and synthesis here, it is immediately apparent that current systems in the various streams regions as well as in the stream as a whole are extremely complex. Actually, these physical aspects of stream dynamics have received little extensive study in the field under natural conditions. A greater amount of study has been carried on in hydraulics laboratories, and considerable theoretical work has been accomplished. For further information, textbooks in hydraulics, and professional publications such as those of the United States Geological Survey, should be consulted.

CURRENTS IN ESTUARIES

The three most important factors operating to produce currents in estuaries are oceanic tides, stream flow, and wind. The interactions of these forces, particularly the somewhat antagonistic processes of oscillating tides (their vertical ranges and lengthwise flow) and unidirectional stream flow (its velocity and volume), serve to make the estuary a restless and complex system of water movements.

Additionally, the morphology of the basin of the estuary and the channel of the stream modify and determine the stream and tidal dynamics. It should be borne in mind that these forces are not regular and constant. Stream flow varies seasonally with rainfall, while tide height and movement are correlated with lunar effects and wind. Winds may serve to increase either tidal movement or stream flow, depending upon the direction and intensity of the moving air mass.

Within most estuaries there exists a relatively regular and uniform rate of water transport. This is to say that the volume of estuarine water discharged to the sea through the mouth of the estuary essentially compensates for the amount of fresh water introduced by the stream into the upper region of the estuary. In an estuary where the incoming fresh water is mixed with the estuarine water, i.e., where no stratification of the two exists, considerable time may be required for the discharge of a particular mass of the fresh water. This time interval is called the *flushing time*. It is an average figure used to describe the period during which a quantity of fresh water derived from stream or seepage remains in the estuary. A simplified formula for determining flushing time is:

$$t_F = \left(\frac{S_8 - S_f}{S_8} V_f \right) \frac{1}{v_D}$$

in which S_s and S_f represent the salinity of sea water at the entrance to the estuary and the salinity of the estuarine water, respectively; V_f is the volume of water in the estuary; V_D is the daily volume movement into and out of the estuary. Flushing time in Great South Bay, Long Island, through Fire Island inlet is of the order of 48 days, or approximately 96 tidal cycles. Obviously a parcel of fresh water entering the estuary near the mouth will move out faster than one in the upper reaches.

The main point is that, on the average, the linear movement of dissolved substances and suspended materials, including living plankton, is a slow process. In many instances the time is sufficient for biological events, important in maintaining the integrity of the estuarine community, to occur.

In a "mixing estuary" the stream flow is held back and often reversed by a strong flood tide. Because of the forces of the two opposing current systems, the reversed flow up the stream is normally slow, ceasing completely as the peak of the flood is reached. Upon the ebb, the outward tidal flow is strengthened by stream flow, the velocity decreasing after the mid-time of the ebb and the movement coming to rest at dead low water. As we have seen for streams, the cross-sectional velocity of an estuary is not constant; a thread of maximum velocity exists. In estuaries, this thread is usually near the surface in the middle of the water course, the precise position depending, however, upon the nature of the shores and bottom.

In many estuaries the typical current pattern is one wherein the lighter, fresh water flows seaward over the upstream movement of denser saline waters (denser by virtue of having a greater concentration of dissolved salts). Under these conditions, a *vertical salinity gradient* exists, and the estuary is said to show stratification. Studies on Chesapeake Bay have indicated that there is a net horizontal seaward flow in the upper layer, and in the lower layer a net horizontal headward flow. The boundary between the two is not level, but slopes toward the right (looking downstream). In the Tamar Estuary of England, it has been found that at some distance upstream both currents may be seen at the surface. During times of considerable fresh-water flow, the more saline waters of the incoming tide move to one side of the basin.

Where stratification and net seaward-headward movements exist, it is possible for certain nutrients and organisms to be transported upstream in the bottom currents, while other materials may be carried to the sea in the upper currents of less saline water. It has been suggested that oyster larvae are moved upstream by lying on the bottom during an ebb tide and then rising into each successive flood tide to be more or less passively transported for the duration of the tide.

In the preceding paragraphs we have seen some of the effects of oceanic tides on the dynamics of the estuary of a stream. In a great number of instances tidal effects are not restricted to the brackish zone, but occur throughout a considerable length of the stream-estuary-sea continuum. Indeed, in the low-lying course of the Amazon, tides serve to bring about measurable variations in stream depth some 960 km above the salt-water zone. The result is the entrance of tides into the stream before the preceding tide has ebbed. In the Amazon there are said to be eight tides along the stream course at one time.

The morphology of the stream channel and estuary basin regulates, to a great extent, the velocity and magnitude of tidal currents. In a wide, unrestricted estuary, currents are typically slow, without significant turbulence. In a narrow, deep basin, rapid, turbulent currents generally occur. If the mouth of the estuary is restricted by depositional features or land closures, the incoming tide may be held back until it suddenly breaks forth into the basin as a tidal wave, or *bore*. In the Severn Estuary,

England, the tidal range may approach 15 m. A bore of 3 m (sometimes reaching 7.5 m) has been reported in the Tsientang River of China. Such currents exert profound effects on the nature of the substrate, turbidity, and biota of the estuary.

Seiches, similar to those described in lakes, may occur in coastal bays and estuaries. The period of these long stationary current systems is determined by the depth and horizontal parameters of the basin.

Oscillations are apparently common in many coastal waters, the effects often obscured, however, by tides. A trinodal oscillation has been recognized in San Francisco Bay, and the great water-level fluctuations of the Bay of Fundy result from the synchronization of tide and seiche, the latter having an amplitude of some 15 m.

The causes of seiches in bays and estuaries are not completely known. It is probable that forces such as variations in atmospheric pressure and wind can supply sufficient energy for the development of standing waves in coastal bodies. As in lakes, friction operates to damp out the oscillation.

The importance of waves and wave action in the estuary is primarily a function of the morphology of the estuary basin. Wherever the mouth of the estuary is restricted and surface area of the body of water slight, there is little opportunity for the development of waves of any appreciable magnitude. Consequently, the role of wave action in watermixing and erosion is negligible. On the other hand, broad estuaries having wide mouths are subject to receiving the full effect of oceanic waves in addition to local disturbances.

Generally, however, the work of moderate Winds are also influential agents in producing certain currents in estuaries. We have already considered the possible role of wind in the development of seiches. Wind contributes greatly to the unusually high waters moved into coastal features during hurricanes and other storms. Depending upon duration and intensity, wind storms may temporarily disrupt the normal circulation patterns in estuaries. Similarly, as the wind is directed, the rate of ebb and flood of tides may be influenced.

GASES IN LAKE, STREAM, AND ESTUARINE WATERS

Many extraordinary properties of water, probably none contribute more to maintaining life in aquatic communities than the capacity of water to hold substances in solution and the ability to enter into chemical reactions. These characteristics result largely from the rather weak nature of the hydrogen bonds linking water molecules.

A relatively small amount of energy is required to separate or to unite the molecules. Therefore, natural waters, including rain, always contain some quantities of dissolved gases and inorganic and organic substances. To repeat an oft-stated truism, "pure water does not occur in nature." On this fact rests the physical-chemicalbiological structure of natural bodies of water that maintains ecosystem continuity as well as evolutionary mechanisms at the organismal level.

The naturally occurring substances that contribute to the "impurity" of waters are derived from various sources, are present in varying states and quantities, undergo certain transformations, and contribute differently to the over-all metabolism of the water community. Certain gases in proper proportion are essential to respiration and photosynthesis in aquatic situations; other gases are lethal to life.

Certain dissolved mineral salts serve as nutrients for free-floating plants; other salts may limit life through osmotic effects. Dissolved organic materials are also present in natural waters, but their reactions are not well known. The nature, quantity, and transformation of substances dissolved in natural bodies of water are generally indicative of the origin of the basin or channel, the climatic regime of the area, and the composition of the substrate in the drainage system.

In this and the following chapter we propose to consider certain properties and reactions of some of the more conspicuous substances contributing to the chemical nature of natural waters as an environment. Although our approach is that of naming the materials and considering them in turn, it is most important to keep in mind the fact that the lake or stream is a dynamic system of chemical interdependencies and interrelationships in association with physical and biological features.

DISSOLVED GASES IN LAKES OXYGEN

Of all the chemical substances in natural waters, oxygen is one of the most significant. It is significant both as a regulator of metabolic processes of community and organism, and as an indicator of lake conditions. Hutchinson has succinctly and aptly stated the case for oxygen in saying: "A skillful limnologist can probably learn more about the nature of a lake from a series of oxygen determinations than from any other kind of chemical data.

If these oxygen determinations are accompanied by observations on Secchi disk transparency, lake colour, and some morphometric data, a very great deal is known about the lake."

The oxygen available for metabolic relationships in natural waters is the oxygen held in simple solution. This is not, as some beginning students are given to think, the O in H_2O. The volume of oxygen dissolved in water at any given time is dependent upon:

- The temperature of the water,
- The partial pressure of the gas in the atmosphere in contact with the water,
- The concentration of dissolved salts (salinity) in the water.

The solubility of oxygen in water is increased by lowering the temperature. For example, the solubility increases about 40 per cent as fresh water cools from 25°C to freezing. Place a ruler or other straightedge across the figure to connect a chosen temperature on the uppermost scale with a given per cent saturation on the middle, inclined scale, say 10°C and 100 per cent saturation. Now read the lowermost scale of oxygen concentration in cc/l at the point where the ruler intersects the scale. Interpreted in one fashion we see that 7.7 cc/l of oxygen constitute 100 per cent saturation at 10°C.

Now connect 5° C and 100 per cent saturation and note that under these conditions a greater concentration of oxygen is dissolved at 100 per cent saturation, 8.7 cc/l to be exact. In other words, solubility increases with decrease in temperature. The original purpose of the nomogram is to determine per cent saturation of a given oxygen concentration at various temperatures and altitudes. Altitude must be considered in order to take pressure into account.

At a given temperature the concentration of a saturated solution of a slightly soluble gas that does not unite chemically with the solvent is very nearly directly proportional to the partial pressure of that gas (Henry's Law).

The solubility of oxygen also relates to Dalton *"law of partial pressures"* which states that the total pressure of a mixture of gases is equal to the sum of the pressures exerted by each of the component gases.

In other words, the solubility of each gas is independent of other gases in the mixture. Under similar conditions of pressure and temperature the solubility of oxygen in water is over twice that of nitrogen and about one third that of carbon dioxide.

The effect of the third factor, the concentration of dissolved salts, upon the solubility of oxygen is that of decreasing the oxygen concentration as salinity increases. At 0°C fresh water at saturation contains slightly over 2 cc/l more oxygen than does average sea water (35 ‰ salinity); at 15°C the difference is about 1.5 cc/l.

Having given attention to the major factors responsible in regulating the solution of oxygen in waters, let us now consider the sources of this gas. Obviously the atmosphere in contact with the lake surface is an unlimited source of oxygen. The volume per cent of oxygen in the atmosphere is calculated to be 20.99, or approximately 210 cc of oxygen per liter of air. This is some 25 times the concentration of oxygen in the same volume of fresh water.

The rate at which atmospheric oxygen passes across the air-water interface and becomes dissolved in the water is dependent upon a number of factors. For one thing, increased wave action or other disturbances at the lake surface results in greater passage of the gas into solution. Secondly, the greater the difference in partial pressure between air and water, the greater the rate of solution.

Thirdly, the less the moisture content of the gas, the more rapid the solution of that gas. Bear in mind, however, that in all of these processes there may be a "two-way" movement, that gas can be lost from the lake to the atmosphere. The *direction* of movement, as well as the *rate,* is determined by the foregoing factors.

Oxygen in natural waters may also be derived from photosynthetic activity. In shallow, nonstratified ponds lacking significant wave action, oxygen may be derived mainly as a by-product of carbohydrate synthesis by rooted plants and by phytoplankton, viz.:

$$6CO_2 + 12H_2O \rightarrow C_6\,H_{12}\,O6 + 6O_2 + 6H_2\,O$$

or mathematically reduced:

$$6CO_2 + 6H_2O \rightarrow C_6\,H_{12}\,O6 + 6O_2$$

Photosynthesis is also a source of dissolved oxygen in large deep lakes. but, in contrast to ponds, this process is restricted to a certain region of the lake.

This zone is delimited by the vertical range of transmission of light effective in photosynthesis. This lighted region is called the *euphotic zone.*

It extends horizontally from shore to shore and vertically from the surface to a level beyond which photosynthesis-effective light fails to penetrate.

As we have seen, turbidity, colour, and the absorptive effect of water itself, serve to quench light; thus they essentially determine the euphotic zone.

In the shallow shore region (the *littoral*), submerged rooted plants along, with phytoplankton contribute oxygen to the lake. In the euphotic zone of the open lake (the *limnetic*), phytoplankton contributes the auto-chthonous oxygen. In most temperate lakes during summer thermal stratification the euphotic zone corresponds closely with the epilimnion.

This accounts mainly for the low-oxygen conditions in the hypolimnion as briefly considered previously. Within the near-surface region of maximum photosynthesis and oxygen gain from the atmosphere, the water often becomes supersaturated with oxygen at the height of diurnal production cycles.

Maximum oxygen production usually occurs in the afternoon on clear days, the minimum immediately after dawn. Cyclic daily fluctuations in 'oxygen concentration, called the *oxygen pulse,* have been observed when conditions are proper.

The lower limit of the cuphotic zone is marked by a level at which organic respiration and decomposition consume oxygen at a rate equal to that at which oxygen is produced over a 24-hr period.

This is the *compensation level*; it will enter significantly into our discussion of productivity.

Below the compensation level in our typical lake lies the *aphotic zone,* a zone in which there is insufficient light to maintain oxygen production at compensation. This is the region of low oxygen content, often depleted. It frequently corresponds with the hypolimnion.

A vertical oxygen distribution pattern such as we have been considering is characteristic of moderately productive lakes of small size.

These lakes, termed *eutrophic* ("rich food"), typically exhibit a hypolimnetic loss of oxygen, the curve of the oxygen distribution dropping sharply through the metalimnion and described as *clinograde.*

It seems generally agreed that temperature and oxidation of organic materials are the major factors in creating the oxygen deficiency in the hypolimnion.

The rate and magnitude of this oxidative breakdown are dependent upon the volume of organic substance supplied to the process.

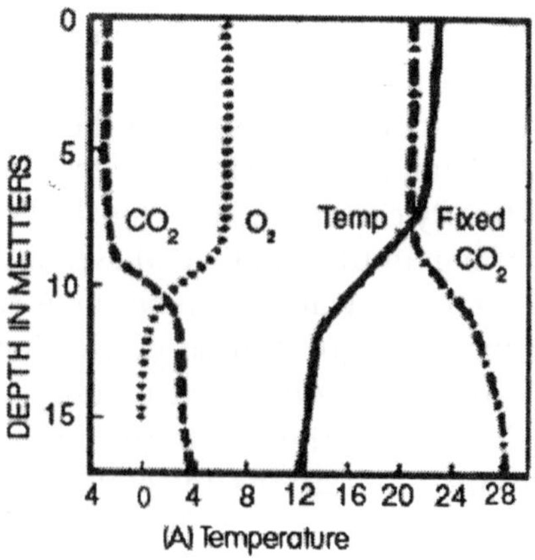

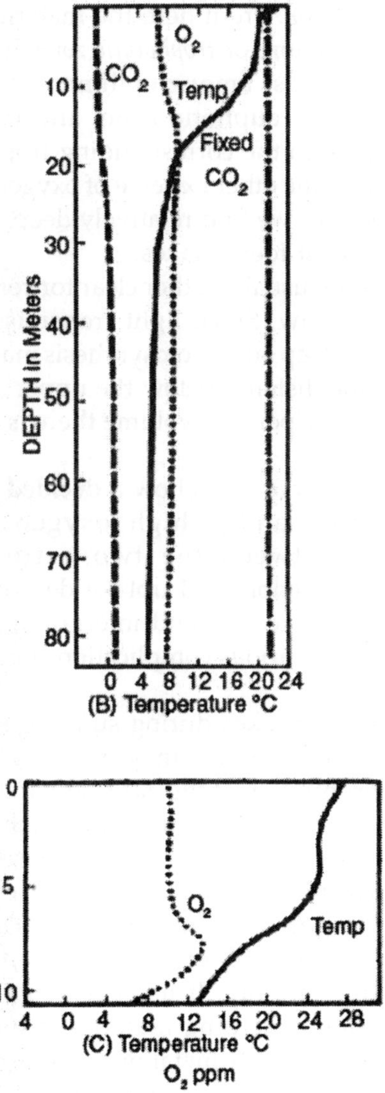

Fig. Various Forms of Oxygen Curves in Lakes. (A) clinograde curve of vertical oxygen; (B) orthograde curve (C) orthograde curve with decrease in oxygen near the bottom.

This substance for the most part is produced in the upper

lighted zone, which, in relation the organic synthesis, is called the *trophogenic region*. From here the materials settle into the zone of decomposition, or *tropholytic zone*. It is now apparent that correlations exist among epilimnetic circulation, light transmission in the euphotic zone, and the production of organic substance in the corresponding trophogenic region.

At somewhat the other extreme of oxygen distribution and biological processes we find relatively deep, unproductive, or oligotrophic ("scant food") lakes.

Such lakes are usually rather clear for considerable depths and turbidity is low. Since light transmission is high, the euphotic zone is deep and photosynthesis may occur, although reduced, at some distance below the upper strata. As a result of low productivity per unit volume there is little oxidation in the hypolimnion.

The oxygen curve may show a decided increase through the metalimnion and a high oxygen content in the hypolimnion. Between the two extremes in oxygen distribution, production, and uptake described here exists a vast range of intermediate and intergrading patterns, many of them exhibiting various complications in oxygen states and metabolism.

In certain clear lakes during summer stratification very high oxygen concentrations may occur in the metalimnion. This great concentration of the gas is apparently derived from levels in and below the epilimnion in which the transparency of water permits a high rate of photosynthesis. The *metalimnetic oxygen maximum* may be followed by a corresponding marked reduction of oxygen in the metalimnion. The mechanism, or mechanisms, responsible for this *oxygen minimum* cannot be described with certainty. It has been suggested that rapid oxidation of slowly settling materials from the epilimnion might account for the low metalimnetic oxygen concentration. Such a process would necessitate that the position of the metalimnion be below the trophogenic zone or at least that photosynthesis proceed at a reduced rate.

Note that underneath the ice in March the concentration of dissolved oxygen is 14 ppm. This saturation was acquired during autumn circulation accompanied by falling temperatures, the latter, of course, increasing the solubility of the gas.

In winter some of the oxygen is consumed by organismal respiration and by organic decomposition. In deep lakes oxygen consumption is relatively small, but in shallow bodies of water the oxygen may become depleted, thereby resulting in mortality of animals.

Vernal circulation redistributes the oxygen as shown for late April and early May. With the development of summer stagnation the hypolimnion becomes quite poor, often completely lacking in oxygen.

In meromictic lakes the oxygen distribution reflects the essential two-region structure of the water mass. Depending upon the circulation pattern above the chemocline, the oxygen content may take the form of a clinograde curve. The waters of the monimolimnion are usually anaerobic. Lakes of the tropical and subtropical regions exhibit widely diverse oxygen relationships.

Polymictic lakes by virtue of continuous circulation, exhibit a generally high, uniformly distributed, oxygen content throughout the year. Oligomictic waters, in which circulation seldom occurs, are typically characterized by a condition of hypolimnetic oxygen depletion. In shallow lakes of the colder regions, for example Alaska, oxygen at near saturation appears to be uniformly distributed by circulation from surface to bottom.

The transport of dissolved oxygen throughout a lake is accomplished primarily by currents set up within the lake and by eddy conduction. Lake currents, as we have previously seen, may be set into operation by wind action on the surface and by density differences between layers in the lake. Circulation in the epilimnion during summer distributes oxygen within that zone.

Eddy systems thrown up along the interfaces between thermal layers or along stream density currents moving through the lake serve to move substances across boundaries, usually at right angles to the current. During seasonal overturn periods the sinking of upper, dense water masses also delivers oxygen to the underlying regions, thereby accounting for the homogeneous oxygen content during those times of uniform density. Simple diffusion of oxygen molecules plays a minor role, indeed, in oxygenation of natural waters. The process is very slow. As a matter of fact, it has been shown that to raise the oxygen concentration 0.4 mg/l at a depth of 10 m by diffusion from the surface would require over 600 years.

With respect to the utilization and ultimate fate of dissolved oxygen in lakes, many aspects have already been considered. In summary, we might say that the major processes acting to consume oxygen are animal and plant respiration, and organic decomposition. During long periods of cloudiness or where waters may otherwise be shaded, for example underneath broad floating mats of vegetation, organismal respiration and decomposition may rapidly take up oxygen. Where decay bacteria occur in great quantity, as in lakes or streams highly polluted with organic sewage, the water often becomes completely anaerobic.

Although we have been concerned primarily in this section with the maintenance of oxygen in lake waters, the source of the gas, and its distribution and fate, we have had to venture into the subject of productivity. This is a natural consequence because there is an obvious direct relationship between the dissolved oxygen content of natural waters and the amount and rate of energy fixation. Recall that the form of the curve of vertical oxygen distribution indicates a great deal about photosynthesis and decomposition at various levels within a single lake, as well as providing a point of comparison of several lakes.

REDOX POTENTIAL

It has been known in chemistry for many years that oxygen plays an important role in chemical changes in various kinds of solutions. Only within the past 30 or so years, however, has investigation of certain oxygen-related phenomena been transferred from laboratory beaker and battery jar to natural waters of lakes and seas. One aspect of these laboratory and field investigations has been concerned with oxidation and reduction processes. This line of research has already been fruitful in contributing to our knowledge of chemical limnology and its influence on the activities and distribution of certain organisms.

The term *oxidation* is applied to the process in which oxygen is added to a substance, or in which hydrogen is lost from a compound, or in which an element loses electrons. Conversely, the loss of oxygen, the addition of hydrogen, or the gain of electrons is termed *reduction*. At the elemental level an electron transfer in which *ferrous* iron is oxidized to ferric iron may be stated:

$$Fe^{++} \rightarrow Fe^{+++} + e$$

Under certain conditions, the reduced iron may undergo a change back to the oxidized state by a shift in electrons as suggested in the reversibility of the process shown above.

The extent to which a substance can undergo oxidation-reduction processes is dependent upon the concentration of other oxidizing-reducing systems and their products in the solution. Within a given solution, the proportion of oxidized to reduced components of a particular system in relation to other systems constitutes the oxidation-reduction potential, or *redox potential*. In a solution, for example lake water, a system with a given redox potential undergoes reduction and oxidizes a system of lower redox value. It may also cause a reduction to take place in a system of higher redox potential.

In practice, the redox potential is determined by immersing a nonreactive electrode (bright platinum is often used) in a solution. The second electrode, necessary to complete the circuit, is usually the hydrogen electrode. The presence of the electrode sets up an electron flow, the direction of which depends upon the proportion of oxidized to reduced material. Where an excess of the reduced state (Fe^{++}) exists, electrons flow to the electrode and oxidation in the system results. In a solution containing a surplus of the oxidized state (Fe^{+++}), flow from the electrode contributes electrons and reduction occurs. The excess or deficit of electrons (relative) can be measured in volts on a potentiometer as an electromotive force about the electrode.

This measure gives the intensity of the force (E_h), either positive or negative, and is the redox potential. A positive E_h reading results from a state tending toward oxidation; a negative E_h indicates a system causing reduction. In addition to *intensity*, the oxidationreduction dynamics also has the attribute of *capacity*. The capacity of an oxidation-reduction system refers to the ability of a system to undergo a certain amount of oxidation-reduction transformation without an intensity change.

Oxygen in natural waters produces a redox potential which is influenced considerably by temperature and the hydrogen ion concentration (pH). Out of consideration for pH effects, the potential is measured at the prevailing pH and then referred to pH 7, this measurement being called the *E*7. Correlations between ranges of E_h at pH 7 and oxidationreduction reactions in certain systems have been established as follows:

NO_3^- to NO_2^-: 0.45 to 0.40 v
NO_2^- to NH_3: 0.40 to 0.35 v
Fe^{+++} to Fe^{++}: 0.30 to 0.20 v
$SO_4^{=}$ to $S^{=}$: 0.10 to 0.06 v

At a temperature of 25°C and pH 7, well-aerated lake waters exhibit a redox potential of about 0.5 v. This potential remains relatively steady as long as the oxygen content is above approximately 1 mg/l. In other words, the redox potential, as such, is affected but little by the oxygen concentration.

Generally speaking, the curve of vertical distribution of the E_h, in lakes follows that of dissolved oxygen and may be a "mirror image" of the ferrous iron. In stratified lakes exhibiting a somewhat orthograde curve of oxygen, the redox potential gives essentially the same pattern. In small lakes in which the oxygen in the hypolimnion is quite low, therefore giving a clinograde oxygen curve, the E_h usually, but not always, appears in a clinograde distribution. Gradations between these forms are common.

Oxidation-reduction activities at the water-mud interface of the lake bottom bear markedly upon the lake chemistry, particularly in the deep hypolimnion, and upon the type of organisms present in the shallow part of the sediments. The E_h of mud exposed to oxygenated water varies near 0.5 v. This potential may extend for several millimeters through an oxidized zone of brownish mud at the surface of the bottom sediments, the *oxidized microzone.* Accompanying the decrease in hypolimnetic oxygen with summer stagnation is a diminution in the depth of the oxidized microzone. As the E_h of the interface approaches 0.2 v, the oxidized microzone may disappear. Below the microzone, the sediments of deep-water lakes are usually highly reducing in nature, the E_h being near zero volt. An oxidized microzone may be persistent in lakes of size and depth sufficient to exhibit an orthograde oxygen distribution.

The processes responsible for maintaining the integrity of the microzone are not completely understood. It was early suggested that molecular diffusion of oxygen through the mud depended upon the reducing state of the sediment. Turbulence may also be important.

We are agreed that low oxidation-reduction potentials suggest the presence of reducing substances which would, in all probability, utilize such free oxygen as might be brought into the solution. For organisms such as anaerobic bacteria, or other organisms not requiring free oxygen for respiration, a low E_h would pose no problem. Anaerobic bacteria exist where the E_h lies below -0.4 v. Oxygen-dependent plants and animals would be restricted from such a zone, although they may be found inhabiting places where the Eh is as low as -0.2 v. Zonation of certain insect larvae in lakes has been correlated with E_h. The midge *Calopsectra* (*Tanytarsus*) sp. dominated the insect fauna in bottom muds over which the Eh of the water was 0.4 or above. Another midge, *Tendipes* (*Chironomus*) sp., characterized muds in waters in which the E_h was below 0.3 v.

CARBONDIOXIDE

The very great importance of carbon dioxide as a contributor to the fitness of natural waters as environment derives from essentially three factors. In the first place it serves in a more or less purely chemical sense to "buffer" the environment against rapid shifts in acidity-alkalinity states. In this sense carbon dioxide ameliorates the chemical environment through the ability of the gas to combine with water to form an acid, and to react to give a neutral salt, or a base. We shall give attention to these reactions later.

A second contribution of importance by carbon dioxide pertains to regulating biological processes in aquatic communities. Seed germination of some plants, as well as plant growth, is determined by the concentration of carbon dioxide. Various animal processes such as respiration and oxygen transport in blood are related to carbon dioxide. A third and most important contribution by carbon dioxide lies in the fact that it contains carbon.

Carbon is one of the most versatile of all elements, due to the possession of four electrons in the outer ring which give great bonding capacity to form a fabulous number of compounds, many of exceeding complexity. The ability to form many compounds is due largely to the asymmetrical nature of the bonds and to the fact that carbon can form chains of atoms almost without limit. This latter characteristic distinguishes carbon from other elements and from inorganic substances, although many inorganic compounds, of course, also contain carbon. Carbon dioxide and water supply the carbon, hydrogen, and oxygen which are major components of protoplasm.

The numerous and varied activities of carbon dioxide in the aquatic ecosystem are made possible primarily by the very high solubility of the gas in natural waters. The solubility of carbon dioxide varies inversely with temperature; at temperatures common in nature, carbon dioxide is much more soluble than oxygen in water. At 20°C and atmospheric pressure of 760 mm Hg, water in equilibrium with atmospheric carbon dioxide contains about 0.88 vol of the gas; only 0.031 vol of oxygen are contained in water under similar conditions. Although air contains some 700 times more oxygen than carbon dioxide (by volume), the proportion in water at equilibrium is more nearly equal, about 4 cc carbon dioxide per liter to 6 cc oxygen per liter. If we consider the amount of carbon dioxide "locked up" in various combined forms, the total amount of the compound in water, particularly the oceans, is much greater than in air.

Carbon dioxide in natural waters is derived from a number of sources. Bacterial decomposition of organic matter in the tropholytic zones, and respiration by animals and plants contribute to the store of carbon dioxide. In the case of plants the greater net contribution is at night, when photosynthesis is not occurring.

Ground waters flowing or seeping into lakes and streams may carry carbon dioxide, the amount being determined by the extent of decomposition in the topsoil and, as we shall presently see, by the chemical nature of the underlying rocks. Within the body of water, certain chemical reactions between acids and various compounds of carbonates release carbon dioxide. Finally, the atmosphere directly furnishes some carbon dioxide to natural waters, and rain, as it falls through the atmosphere, dissolves some of the gas and delivers it to lakes and other waters.

The importance of rain in supplying carbon dioxide to inland waters lies in reactions wherein carbon dioxide is maintained and transported in forms other than as a gas. As rain percolates through soil containing carbon dioxide of decomposition, some of the gas becomes dissolved in rain water. The reaction between CO_2 and H_2O results in the formation of carbonic acid (H_2CO_3). If this weak acid encounters carbonate-holding rocks, limestone ($CaCO_3$) for example, the latter dissolves as calcium bicarbonate, or $Ca(HCO_3)_2$. The solution of $Ca(HCO_3)_2$ remains stable only in the presence of a certain amount of *free* or *equilibrium, carbon dioxide*. Free carbon dioxide represents the CO_2 in H_2CO_3 plus that in simple solution. The formation of H_2CO_3 and its dissociations are shown in the reactions:

$$CO_2 + H_2O \rightarrow H_2CO_3 \rightarrow H^+ + HCO_3^- \rightarrow H^+ + CO^2_3 \quad (I)$$

Note in the above reactions that carbon dioxide is contained in two states not previously encountered, i.e., as bicarbonate and carbonate radicals, HCO_3^- and CO_3^{2-}, respectively. This carbon dioxide is called the *combined carbon dioxide*. Solution of $CaCO_3$ is dependent upon the addition of CO_2 in an amount greater than that of free carbon dioxide; this additional CO_2 is known as *aggressive carbon dioxide*. The major points to be gained from these considerations include:

- The relations between rain and soil in supplying compounds containing carbon dioxide to natural waters,
- The chemical reactions by which those compounds are formed,
- The occurrence of carbon dioxide in its three forms: free (CO_2 in solution plus that in H_2CO_3), half bound (HCO_3), and bound (CO_3^{2-}).

The direction of reactions involving the forms of carbon dioxide and the very occurrence of these in natural waters are reciprocally related to acid-base relationships in the medium. When carbon dioxide dissolves, the reaction and end products depend upon the nature of the solvent, particularly with relation to the hydrogen-ion concentration. Under acid conditions, the combination is as shown in Equation above in which H_2CO_3 is formed followed by dissociation to $H^+ + HCO_3^-$.

Under highly basic conditions the reaction is:

$$\text{Base OH} + H_2CO_3 = H_2O + \text{base}^+ + HCO_3^-$$

CHEMICAL BUFFERING

The maintenance of near-neutral conditions in mineralized waters is due to *buffering* by chemical systems such as the carbon dioxide-bicarbonate carbonate complex. Other systems may involve magnesium, sodium, or potassium. In other words, as acid conditions arise, the reaction between acid and base from the bound carbonate, for example, brings about an increase in neutral bicarbonate. Continued acidification releases carbon dioxide and carbonic acid from the bicarbonate accompanied by loss of carbon dioxide from the system.

This reaction is the basis for certain limnological techniques for determining the so-called *alkalinity* of natural waters. In these tests a quantity of strong acid is added to water in the presence of a proper indicator. The amount of acid necessary to convert any carbonate or bicarbonate present to free CO_2 is a measure of the HCO_3- and CO_3^{2-} in solution. In this sense, alkalinity refers to the quality and quantity of compounds which bring about a shift in the pH of a solution toward the alkaline side of the pH range.

Although not always the case, alkalinity usually reflects the activity of calcium carbonate. Therefore, three forms of alkalinity may be recognized: bicarbonate, normal carbonate (*phenolphthalein alkalinity*), and hydroxide. At the other end, increased alkalinity of the solution brings about a reaction between the base and carbonic acid to hold the departure from neutrality to a smaller value than would otherwise be the case if the buffer system were not present.

DEFINITION OF pH

The pH is the logarithm of the reciprocal of the hydrogen-ion (or more properly, the hydronium-ion) activity. The pH may be expressed mathematically as follows:

$$pH = \log \frac{1}{(H^+)}$$

where (H+) is the amount of hydrogen ions in a solution in moles per liter. In a liter of pure water there is 0.0000001 of hydrogen mole/ions (and a corresponding quantity of (OH-)). The pH of pure (neutral) water is, therefore,

$$pH = \log \frac{1}{0.0000001}$$

Increase in the concentration of H+ ions results in a lower pH value, or conversely, reduction in the H+ concentration brings about a higher value. From this relationship, a pH scale has been devised. This scale ranges from pH 0, corresponding to a solution with $(H^+) = 1$, through pH 7 (or neutrality) to pH 14, corresponding to a solution with $(H^+) = 10^{-14}$. From pH 0 to pH 7 solutions are *acid*. From pH 7 to pH 14 reactions are *alkaline*. With respect to the buffering system, we note that only free carbon dioxide is of any import in natural water systems below pH 5, that bicarbonate dominates the range from pH 7 to pH 9, and that the carbonate radical is most important in the range above pH 9.5 or 10.

The pH range of lakes having some degree of flow through the basin is generally from about 6 to 9. In limestone regions the dissolved carbonates may extend the pH range considerably beyond 9. In basins lacking outlets, evaporation may concentrate alkaline substances, resulting in pH readings of over 12. At the other extreme, accumulation of acids such as sulfuric acid in volcanic lakes gives a pH as low as 1.7.

OCCURRENCE OF CARBON DIOXIDE IN LAKES

Having given attention to some of the major aspects of the sources of carbon dioxide in natural waters, as well as to the complex and often critically balanced reactions of the forms of the compound, let us now turn our attention to general considerations of the compound in lakes.

From what we have seen of the close relationship between the chemical nature of the drainage basin substrate and the chemistry of waters of the basin, we should expect wide regional variation in carbon dioxide content of lakes. Newly formed lakes in regions of weakly soluble rocks typically contain little carbon dioxide in any form.

In these situations the low quantity of carbonates in the substrate results in little of the bound or half-bound forms of carbon dioxide. Similarly, the paucity of soluble minerals as nutrients in biological processes inhibits the development of large biotic populations that would contribute carbon dioxide through respiration, and, of course, decomposition is reduced. Lakes of this type are usually slightly acid, the pH ranging near 6. Lakes of higher acidity (pH 4 to 6) are common in regions of lowlands and bogs. In these waters the free carbon dioxide content is usually quite high, ranging to nearly 200 ppm.

As would be expected in view of the low pH, the concentration of bound carbon dioxide as carbonate is low, usually less than 9 or 10 ppm. Lakes having these characteristics are termed *soft-water lakes.* A large number of lakes fall into a class in which the pH is circumneutral, and which may be called *medium-water lakes*. Free gaseous carbon dioxide in medium-water lakes varies widely, frequently showing supersaturation relative to the partial pressure of the gas in the atmosphere.

These lakes may contain bound carbon dioxide up to 30 or 35 ppm. In regions where the substrate contains easily dissolved minerals, *hard-water lakes* occur. These lakes are characterized by negative values for free carbon dioxide due to of bicarbonates at a greater rate than carbonates are precipitated, and by pH values ranging from about 8.5 upwards. Bound carbon dioxide content amounts to over 35 or 40 ppm, often reaching 200 ppm or more.

SEASONAL CYCLE OF CARBON DIOXIDE AND pH

During winter in the colder regions ice cover forms over the lake, thereby inhibiting exchange of materials across the air-water interface. Thus the lake becomes essentially "sealed in." Oxidation of organic substances, particularly in the depths, consumes oxygen and increases the carbon dioxide content. This often results in the building up of a gradient of metabolic substances from surface to bottom, especially if there is some transmission of light through the ice permitting photosynthesis in the upper, unfrozen waters.

Under such conditions the waters immediately below the ice may contain a considerable quantity of oxygen, sometimes reaching near saturation, and little or no free carbon dioxide. The pH in this zone may show slightly alkaline conditions with low bicarbonate concentration and absence of carbonates. Remember that we should expect to find carbonate only in the absence of free carbon dioxide. In the deep waters, bicarbonates and free carbon dioxide are increased, while oxygen may be nearly or completely depleted; the pH may drop as during summer stratification.

In the warmer regions of the temperate zones, nonfreezing lakes typically contain quantities of carbon dioxide more or less uniformly distributed throughout the waters during winter. This is, of course, due to circulation during the cold season. At this time the phenophthalein alkalinity of surface waters is usually nil, and carbonates are absent.

In holomictic lakes the period of vernal mixing brings about the relatively even distribution of all dissolved materials. During this overturn the pH of the water is uniform from surface to bottom.

Free carbon dioxide, together with other gases such as methane and hydrogen sulfide, derived from winter decomposition processes, is lost to the atmosphere. Following the loss of the free form, bound carbon dioxide reappears as carbonate in medium and hard-water lakes as summer stratification begins.

We have already seen that in productive holomictic lakes exhibiting two circulation periods the oxygen concentration diminishes rapidly with depth during summer, giving a clinograde curve.

In these lakes the carbon dioxide and carbonate concentration show a general inverse relationship to the oxygen; that is, the concentration of carbon dioxide and bicarbonate increases slightly with depth. On the other hand, an orthograde oxygen distribution is usually acompanied by only slight increase, if any, in carbon dioxide.

During the summer the epilimnion of medium and hard-water lakes is typically devoid of free carbon dioxide, and contains measurable quantities of carbonates. It has been shown, for example, that in Douglas Lake, Michigan, in July, free carbon dioxide may be absent to a depth of about 14 m, and in this same distance carbonates range from about 10 to 8 ppm. pH in the upper waters is slightly over 8. At a depth of 20 m, the free carbon dioxide content reaches 11 ppm at a pH of 7.1. As a point of interest, records over a 30-yr period show that the surface alkalinity due to half-bound carbon dioxide varied only from 110 to 128 ppm.

Autumnal circulation brings about uniform physical and chemical conditions throughout the lake. Vertical gradients in density, temperature, dissolved gases, and solids, which became established during summer stagnation, are broken down. The hypolimnetic gases of decomposition are thrown off at the lake surface. The lake's oxygen content is circulated, and with lowering temperatures of autumn an increased supply is dissolved in the water.

OTHER LAKE GASES

Methane is an organic gas (CH_4) widely called "marsh gas," common in many alkaline lakes, ponds, and swamps, particularly during summer stratification. It is produced by bacterial decomposition of organic substances in the tropholytic zone, primarily in the lake bottom.

The process involves a multistage breakdown of complex organic material to compounds of simple molecular structure and then decomposition of these, releasing methane and carbon dioxide. It occurs only under anaerobic conditions. In carbohydrates, for example, cellulose is attacked by bacterial enzymes and hydrolyzed to a simple sugar. Anaerobic decomposition of the sugars may result in the formation of hydrogen and methane. The time of highest production appears to be during summer stagnation when the bottom muds have become significantly reducing, that is, when the oxidation potential is low. The process can proceed at relatively low, temperatures, about 5°C, because of the tolerance of at least one of the methaneproducing bacteria.

In shallow bodies of water such as ponds and swamps, and in the shore zones of lakes, bubbles containing methane and other gases are often seen rising to the surface and erupting. The formation of bubbles is apparently ssdue to insufficient water in the shallow situations to dissolve the gas as it is formed in bottom mud and debris. It follows then that the occurrence of bubbles at the surface over deep water would be unlikely. Under winter conditions methane bubbles may be present in the ice cover.

Analysis of the gases in bubbles in a Russian lake (Beloye) revealed a methane proportion of from 74 to about 84 per cent, and 5 to 18 per cent hydrogen. As the bubbles rise to the surface, the hydrogen is lost and the methane volume diminishes to near 24 per cent. In the ascent, nitrogen and a small quantity of oxygen are gained. These transfers of gases into and out of the bubble operate, of course, under the law of partial pressures. Although the full story of methane and its role is not known, it does appear that some of the gas is oxidized by organisms in oxygenated zones of lakes as the bubbles rise. Here we see yet another phase in lake metabolism serving to decrease the oxygen content of deep waters.

Additional gases occur in natural waters and should be mentioned at this point, even though certain of them occur in small quantities. These will be considered in relation to dissolved solids. As we have already seen, hydrogen is formed as a decomposition product in the anaerobic zones of bodies of water. It acts, in part, to form methane and also occurs free in bubbles.

Free ammonia may, under certain conditions, be present in small quantities in lakes and streams, and elemental nitrogen, derived mainly from the atmosphere, is highly important in lake metabolism; these substances will be taken up with nitrates. Hydrogen sulfide, a decomposition product, is frequently present in the hypolimnion of certain lakes during summer; it will be considered in the following chapter along with sulfur.

DISSOLVED GASES IN STREAMS

The behaviour and basic relations of gases in streams follow the same fundamental physical and biochemical laws as operate in lentic situations. However, temporal and spatial relationships of dissolved substances, generally, are variously modified and complicated by the very features that characterize streams.

In other words, the presence of a current with its inherent turbulence effects, the considerable exchange between stream and surrounding terrain, and the variations in water volume and chemistry associated with climate and drainage basin morphology all serve to make a momentary or long-term picture of stream conditions quite different from a lake.

During our time another factor, an "unnatural" one, has become important in relationships in streams. This is pollution, and although the act of ravaging our waters by dumping wastes into them is not to be condoned, we have learned much about biochemical reactions from pollution studies. This topic will be studied more fully in a later chapter.

OXYGEN IN STREAMS

There are three primary sources of oxygen in stream water, the contributions of each being far from equal and indeed varying greatly with time of day, season, current velocity and stream morphology, temperature, and biological characteristics.

GROUND WATER AND SURFACE RUNOFF

For most streams these sources are relatively insignificant in supplying oxygen. Water issuing from springs, subterranean channels, or from seepages is typically low in dissolved oxygen, often to the point of being anaerobic. Not only do these ground waters fail to provide oxygen to the spring run, but also the run itself dilute the oxygen content of a parent stream at the point of junction of the two streams.

If, however, subsurface waters flow over broken rocks shortly, before reaching the surface, the waters may be near saturation. Similarly, if surface runoff is rapid and vigorous, the water may be high in óxygen content; but sluggish surface drainage seldom contributes great oxygen stores to running waters.

PHOTOSYNTHESIS

In less turbid streams vegetation contributes oxygen to the waters during the day. Rich growths of algae on submerged objects such as rocks or logs, and algae floating free in the water, together with higher plants growing beneath the surface, produce high amounts of oxygen, particularly on cloudless days. During the night and on cloudy days this production may be somewhat balanced by respiratory consumption of the oxygen by plants and animals.

In the shallow headwater reaches of a stream a net production of oxygen by photosynthesis contributes to the downstream content. In the lower, more turbid regions of most streams, however, local photosynthesis probably contributes little to the downstream oxygen content.

Since turbidity is to a great extent a function of stream discharge and capacity, we can again recognize the importance of stream channel and water-shed features in the chemistry and biology of streams.

PHYSICAL AERATION

The introduction of a large amout of organic substance such as sewage, or debris from swamp or marsh flooding into a stream may bring about a depression of the dissolved oxygen content below the saturation value. The difference between the actual oxygen content and the amount that could be present at saturation is called the *saturation deficit.*

This deficit is incurred, of course, through the uptake of oxygen by aerobic decomposition of the organic materials in the stream. Yet, downstream, barring no further immediate depressions, the stream shows evidence of regaining its earlier level of oxygen concentration. The oxygen serving to offset the oxidative loss is absorbed from the atmosphere' through reaeration of the stream waters. Reaeration is, therefore, a process by which streams secure oxygen directly from the atmosphere, the gas then entering into the biochemical oxidation reactions in the stream.

Within the stream, distribution of the oxygen derived from the atmosphere is accomplished by turbulent transport. The rate at which reoxygenation of a given parcel of water takes place depends upon a number of factors, including temperature, degree of turbulence, depth of the parcel, magnitude of the saturation deficit and, naturally, the momentary oxygen demand by decomposition processes.

The importance of temperature lies in its inverse relationship to oxygen solubility, and to its influence on metabolic demands of organisms. Turbulence is a highly variable factor in the oxygenationdeoxygenation relationship, varying from negligible in quiet pools to highly important in riffles and rapids.How effective is reaeration as a method of oxygenating flowing waters?

Note particularly the relationships between discharge and saturation deficit, and the dissolved oxygen content. The increase in oxygen concentration is greater at the lower discharge than at the higher.

The total mass picked up, however, is greater in the higher discharges.Under natural conditions, the waters of streams typically contain a relatively high concentration of oxygen tending toward saturation.

However, a number of factors operate to varying extents to reduce the oxygen content and to contribute to the loss of the gas from streams.

- *Turbulence*: We have just witnessed the fact that turbulence plays an important part in the aeration of streams. It follows, therefore, that physical aeration is reduced with decrease in turbulent flow.
- In stream areas below riffles and falls, the water is often saturated with oxygen throughout the day and night. In quiet reaches and in streams of low velocity, the waters may be below saturation during the night.
- *Respiration of Organisms*: Respiratory activities of plants and animals, and oxidation of organic matter utilize dissolved oxygen of streams. The effects of these processes are more conspicuous at night, being masked during the day by photosynthesis.
- *Photosynthesis*: In the more stable zones of streams, submerged plants contribute in a major way to the oxygen content of the water. Consequently, fluctuations in photosynthetic activity will be reflected in the amount of dissolved oxygen present.
- *Temperature*: In streams, as in other waters, solubility of oxygen varies inversely with temperature. Thus, raising of water temperature could result in loss of oxygen from streams.
- *Atmospheric Pressure*: Since the solubility of oxygen bears a direct relationship with atmospheric pressure, reduction of pressure would bring about a decrease in the amount of dissolved oxygen.
- *Inorganic Reactions*: Certain inorganic activities, such

as the oxidation of iron, may contribute to the loss of oxygen from polluted streams.

- *Inflow of Tributaries*: The introduction of tributary waters of low oxygen content serves to dilute the concentration of oxygen in the receiving streams. This effect is especially noticeable with the entrance of spring or some seepage waters.

The annual cycle of oxygen of streams is closely correlated with temperature conditions. Studies of large rivers and small streams in warm southern regions of moderate temperature regimes and in northern climates of broad temperature fluctuations have shown that the oxygen content of flowing waters is generally highest in winter and lowest in late summer.

This primary temperatureoxygen relationship may, however, be tempered by a number of factors acting throughout the year. In small, slow streams of northern latitudes, decreased day length and ice and snow cover may serve to inhibit photosynthesis, thereby bringing about a short-term depression of the winter peak of oxygen concentration.

The vernal decline of oxygen content in slow, clear streams may be attributed to the action of spring floods in removing vegetation. Further, decrease in oxygen content toward late summer, in streams generally, may be due to one or more of several factors.

Water temperatures reaching their maxima in late summer hold less of the gas in solution; decreased discharge results in diminished physical mixing and reoxygenation; greater decomposition of summerproduced organic material utilizes some of the available oxygen. On the other hand, a rapid and abundant growth ("bloonm") of phytoplankton often increases the oxygen content during the summer.

A daily oxygen rhythm, the diurnal pulse, is largely a reflection of temperature fluctuations and photosynthesis-respiration relationships. In mountain and hill country the water is typically at between 95 and 105 per cent saturation at all times, so long as there 'is not a dense growth of rooted aquatic plants.

Turbulence is of great importance and easily maintains an average of 100 per cent saturation. In a slow, shallow stream containing a fair abundance of rooted plant growth, daytime photosynthetic production of oxygen exceeds turbulent diffusion into the air and the respiratory consumption of the gas. The result is frequent super saturation with oxygen during the day, followed by a drop to 100 per cent saturation at night. In winter in the temperate zone the time of maximum daily oxygen production corresponds closely with the time of highest temperature.

In summer, oxygen production lags behind rising temperature, often by two or three hours, thus imparting high saturation values even after sundown. Depending upon oxygen demand, the minimum level usually occurs prior to the early-morning temperature low. In streams of this naute the oxygen concentration may be acutely affected by sunlight intensity.

The depression in the diurnal curve between about 1300 and 1500 hr is doubtless due to cloudiness shown for the same periods on the light-intensity curve. Such data as these indicate the importance of insolation and photosnythesis in maintainiiig the oxygen content of slow streams. The considerable diurnal oxygen range in such streams, also points up the importance of more than a single oxygen determination in a stream study.

Obviously an average of several measurements would be much more meaningful than one, and a diel series would most satisfactorily describe the oxygen conditions in this stream type. We have already learned that oxygenation through turbulent mixing is a prominent feature of fast waters and even highly turbid streams of low discharge. The diurnal oxygen pulse of such streams as these is usually less pronounced, since mechanical aeration and oxygen consumption are essentially uniform within a given stream segment.

The distribution of dissolved oxygen throughout the linear extent of a stream is subject to many variables, so much so that generalizations relating to the feature are not readily apparent. As we have seen, the upper reaches of streams are usually well oxygenated, the concentration being a function of turbulence and water temperature.

Farther along the gradient, the stream becomes slower and rich growths of submerged plants may develop. In this region the oxygen content is influenced less by physical aeration and more by organic oxygen production and respiration.

The maximum-minimum relationships here may be the reverse of those found upstream, in that maximum saturation values will be found in the afternoon and the lowest values in the early morning. In the lowlands, increased turbidity and organic decomposition produce a generally low oxygen content.

In clear, slow streams supporting an abundant flora, oxygen may be added to the waters along the stream course, thereby resulting in a net increase in downstream segments. The introduction of organic pollutants and inflow, from marshes and swamps may serve to decrease the oxygen content at and below the point of inflow of these materials.

Thus the amount of dissolved oxygen distributed through the course of a natural stream is strongly determined by channel and flow characteristics and by biochemical interactions and processes.

Although our knowledge of reservoir limnology is meager, we know that the effects of the release of impounded waters into streams below the impoundment are varied and important, and worthy of considerably more research in the future.

For example, we know that release of water from a reservoir through deep-water intakes during summer stratification introduces poorly oxygenated water and organic materials from the hypolimnion into the stream below the impoundment.

Under extreme conditions this actual dilution of the stream and introduction of oxidizable matter could have deleterious effects on the biota. On the other hand, draw-down through shallow intakes from the epilimnion may serve to increase the dissolved oxygen content of the downstream waters.

CARBON DIOXIDE AND PH IN STREAMS

In streams the occurrence and abundance of components of the bicarbonate buffer system and the pH condition are determined primarily by current, biological processes, and the chemical nature of the substrate. The role of current is that of ameliorating the chemical climate of the stream. This is accomplished through mixing and moving concentrations of substances, but usually within a relatively restricted segment of the stream.

Over a considerable distance, and depending upon the volume of introduced materials, the chemical composition of stream waters is subject to considerable change. The biological processes acting to influence the nature of the water are those previously considered, i.e., photosynthesis and respiration. In general, the behaviour of carbon dioxide in streams is similar to that of oxygen, but with inverse properties.

In stream areas of abundant plant growth the concentration of carbon dioxide is minimum during most of the day and maximum in the early morning hours. In sluggish streams a phytoplankton bloom may indeed exhaust the free carbon dioxide supply and obtain further carbon dioxide from bicarbonates. In the lower, sluggish stream course under conditions of high turbidity due to organic suspensoids, high carbon dioxide content occurs in the presence of depressed oxygen concentration, owing to bacterial action.

The chemical composition of mineral-bearing rocks in the stream valley and channel, and also the drainage nature of the valley, may act in a major way to determine the water composition.

Under certain conditions these factors may somewhat offset the influences of biological processes. For example, in limestone regions a high concentration of carbon dioxide is not apt to develop in the stream because, as we have already seen, any excess of gas would enter into combination with line in the substrate to give a carbonate. The direct strong influence of the substrate rests primarily upon the solubility of tile buffer sunstances in rocks. Streams issuing from or flowing over relatively insoluble igneous formations of high silica content are typically soft, because the bicarbonate content is insufficient to buffer pH changes due to accumulation of carbon dioxide.

As a result of a shift of the buffer system toward carbonic acid, the pH of such streams will be below neutrality. However, atmosphere-water equilibrium essentially adjusts the carbon dioxide content in such a way that the pH value does not go far toward the acid side, usually coming to balance at about pH 6.

If, on the other hand, the stream encounters water from bogs, which commonly occur in areas of siliceous formations, the pH may be further lowered.

In many regions, the Appalachian slopes of North America for example, the fate of soft-water streams is to flow eventually over sedimentary formations in the lower reaches. These formations are normally rich in soluble carbonates, and the addition of carbonate ions serves to shift the pH value toward the basic range of the scale, usually to pH 8 or higher. The increase in pH value is accompanied by a marked rise in total alkalinity, and decrease in carbon dioxide.

Highly acid streams occur primarily on low marshy or swampy terrain, on poorly drained sandy "flatwoods," or under special conditions. These streams are usually stained brownish and support a relatively meager biota.

The nature of the acids contained in such streams is not fully known, but frequently considered to be humic, yellow organic, or perhaps sulfuric.

In the southern United States, the pH value in sluggish acid streams may range near pH 4, with a concurrent high concentration of free carbon dioxide (25 ppm or more) and low bicarbonate content. Oxygen in these streams is often low, being on the order of 20 to 30 per cent saturation at 17° to 18°C.

Hard-water streams occur commonly as spring runs or surface drainage streams in regions of soluble basic geologic formations.

Calcium and magnesium carbonate, often occurring together, are prominent sources of ions which, in solution, contribute to the hardness of streams; however, sulfates, chlorides, and other compounds may also contribute to total hardness.

The combination of high carbonate concentration and low carbon dioxide content, characteristic of these streams, apparently constitutes a more favorable environment for plants and animals than do acid conditions-for, generally speaking, hard-water streams support a varied and abundant association of organisms.

Most unpolluted major streams exhibit a pH value on the alkaline side of neutrality, and, indeed, tend toward uniformity of composition generally.

Although the relationships of carbon dioxide, bicarbonates, and carbonates will be considered a bit further with respect to dissolved solids, we might briefly summarize their activities in relation to pH in lakes and streams thus:

- The pH value varies inversely as the dissolved carbon dioxide concentration, and directly as the bicarbonate concentration;
- The critical value relating to the presence or absence of free carbon dioxide is pH 8, the free gas being absent above that value;
- The absence of free carbon dioxide does not limit photosynthesis of certain algae and higher plants, some being adapted for utilization of carbon dioxide from carbonates, usually resulting in very high pH values.

Methane, hydrogen sulfide, and other gases derived from breakdown of organic substance may be present in high concentrations under proper conditions in very slow streams and in stagnant regions of stream pools. In moving water, however, turbulence due to current tends to diffuse and eliminate the gases.

DISSOLVED GASES IN ESTUARIES

The considerable differences between the chemistry of fresh water and that of sea water bring about complex relations of dissolved substances in general within an estuary. The occurrence of dissolved materials relates to the disproportion of dissolved materials found in the waters of the inflowing stream at the upper end of the estuary and in the sea water at the mouth of the estuary. Between these two relatively uniform states there exists a considerable gradient in processes and conditions.

Dissolved gases are distributed in an estuary in accordance with turbulence and current factors, biological activities, and salinity and temperature effects.

We have previously considered most of these factors and their influence on dissolved gases with respect to lakes and streams; in general, the interrelationships are similar in estuaries. One new factor is quite influential in estuaries; this is salinity, the total amount of dissolved inorganic solids.

OXYGEN

In salt water the solubility of oxygen decreases as water temperature and salinity increase. If we visualize cool, fresh (low-salinity) water entering the uppermost reaches of an estuary and grading toward somewhat warmer and more saline sea water at the lower extreme, we can begin to appreciate the adjustments taking place throughout the length of the estuary.

Table gives saturation (absorption) coefficient values of oxygen in water at various temperatures and salinities. The values show not only that temperature is the most important factor in determining oxygen solubility, but also that salinity manifests considerable influence.

We see then that less oxygen can be dissolved in sea water than in fresh water. At saturation at 15°C, a liter of sea water (at 36 ‰ salinity) contains 5.8 cc oxygen per liter of water; a liter of fresh water at the same temperature holds 10.3 cc. The importance of these relationships lies not only in the aforementioned possibility of a considerable linear oxygen gradient in a mixing estuary.

There is also the possibility of fluctuations associated with stream flood seasons and the inflow of large quantities of fresh water, or with dry seasons when tidal flow of sea water dominates.

Table. Saturation Coefficient Values of Oxygen And Carbon Dioxide In Water

Salinity (‰)	*Tempera ture* (°C)					
	0°	12°	24°			
	O_2	CO_2	O_2	CO_2	O_2	CO_2
0 (fresh water)	49.24	1717	36.75	1118	29.38	782
28.9	40.1	1489	30.6	980	24.8	695
36.1 (sea water)	38.0	1438	29.1	947	23.6	677

In nonmixing estuaries salinity stratification during summer often results in conspicuous differences in dissolved oxygen content in deep and in surface waters. In parts of Chesapeake Bay during summer the oxygen concentration may range from 90 to 100 per cent saturation at the surface, while bottom waters show from 40 to 50 per cent saturation. The surfacebottom differential decreases upstream as the depth decreases. In these shallow zones there is probably more mixing than in the bay proper.

The range of oxygen concentration in the bay reflects the activity in the lighted trophogenic zone and the relatively slow replacement of sea water in the lower level. It may well be that even with more rapid replacement the water from the sea might be poor in dissolved oxygen.

Along with vertical differences, oxygen characteristically varies diurnally and seasonally within estuaries. The ranges of such variations differ, depending upon the nature of the freshwater source, the morphology of the basin, and effects of tides. In deep, turbid estuaries lacking the con-

tribution from an abundant bottom flora, diurnal oxygen pulses are apt to be relatively slight. In Chesapeake Bay the range for surface waters was found to be from about 85 per cent of saturation in early morning to about 115 per cent in late afternoon.

Shallow, clear estuaries may contain bottom growths of algae on which oxygen bubbles may be seen. In waters such as these the minimum-maximum diurnal range may exceed 200 per cent. The effects of incoming sea water during a tidal cycle may serve to mask the local conditions. Seasonally, the oxygen dynamics in an estuary may be influenced by variations in river discharge, tides, and day length and biological effects. Surface waters in one part of Chesapeake Bay were found to have 143 per cent saturation with oxygen in April, and about 42 per cent in August. Bottom waters in the same area reached 133 per cent saturation in October, and 24 per cent in June.

THE CARBON DIOXIDE SYSTEM AND pH

The solubility of carbon dioxide in estuarine waters is determined primarily by the amount of sea water mixing with fresh water, and secondarily by temperature. The importance of sea water as a solubility factor rests, of course, upon salinity. In Table we see that at any temperature the absorption coefficient value of carbon dioxide decreases as the salinity increases. The high solubility of carbon dioxide is due to its chemical reaction with water. Although some of the carbon dioxide in sea water is in the form of the free gas and as carbonic acid, much more is present as bicarbonate and carbonates.

This condition results from the fact that sea water contains alkaline radicals in excess of the equivalent acid radicals which, as we have seen, shift the carbon dioxide system toward the carbonate formation, thereby reducing the free carbon dioxide concentration.

The presence of the excess bases—boric acid and its borates, carbonic acid and the carbonates—in sea water serves to buffer the water against great changes in pH that might develop from the addition of acids or bases. The pH of sea water at the surface is very stable, usually ranging between pH 8.1 and 8.3.

Since carbon dioxide uptake is greater in the presence of excess base, and since river waters usually contain a lower concentration of excess bases than sea water, we should expect to find a lower content of free carbon dioxide in the mouth of the estuary than in the upper reaches. Similarly, because river waters are seldom buffered, the free carbon dioxide concentration and pH should be more variable in that part of the estuary dominated by stream influences.

Streams transporting large quantities of humic material in colloidal suspension are frequently slightly acid. Upon meeting sea water the colloidal particles are coagulated and the pH shifts toward the alkaline side of neutrality. East Bay, Texas, receives considerable runoff from organically rich salt marshes, and it was found that during summer the pH ranged from 6.9 throughout much of the bay to 7.8 near the mouth where it discharges into the Gulf of Mexico. Gulf waters during the same period gave a pH value of 8.0.

In view of what we have learned of oxygen-carbon dioxide-pH relationships generally, we should expect the vertical, diurnal, and seasonal distributions of carbon dioxide and pH to operate acording to certain principles. Although few actual studies have been made of carbon dioxide in estuaries, the pattern of distribution is expected to be the reverse of that of oxygen. By the same rules, pH values should vary inversely as the free carbon dioxide content and directly as the dissolved oxygen concentration.

Chapter 4

Dissolved Solids in Natural Waters

Practically all of the naturally occurring elements of the earth's crust could probably be found in inland waters. Some of these substances would be expected to occur in minute concentrations, however.

In addition to the gaseous components considered in the preceding chapter, the more conspicuous materials which are found in varying quantities in natural waters include carbonates, chlorides, sulfates, phosphates, and often nitrates. These anions occur in combination with such metallic cations as calcium, sodium, potassium, magnesium, and iron to form ionizable salts.

As a result of availability and high solubility of carbon dioxide, carbonates are usually the most abundant salts in fresh waters. Both the quantitative and qualitative aspects of the chemical composition of inland waters are influenced to a high degree by the geochemistry of the terrain and, in the case of lakes, especially by the form of the basin with respect to inflow and outflow.

In general, the inorganic composition of the water of an open lake, that is, one with effluents through which water moves, is a reflection of the nature of the influent waters. The composition of these lakes approaches that shown in Table. The inorganic composition of closed lakes (or those lacking significant effluents) is greatly modified by precipitation and concentration of salts as determined by evaporation. Lakes near coasts may receive substances from sea spray.

Table. Mean Percentage Composition Inland Waters

Ion	*Percentage*
CO_3^{2-}	33.40
SO_4^{2-}	15-31
Cl^-	7.44
NO_3^-	1.15
Ca^{2+}	19.36
Mg^{2+}	4.87
Na^+	7.46
K^+	1.77
$(Fe, Al)_2O_3$	0.64
SiO_2	8.60

TOTAL DISSOLVED SOLIDS

The total concentration of dissolved substances or minerals in natural waters is a useful parameter in describing the chemical density as a fitness factor, and as a general measure of edaphic relationships that contribute to productivity within the body of water. One measure called *total dissolved solids* is determined by simply evaporating a filtered quantity of water at low temperatures.

The dried residue contains both inorganic and organic materials. Ignition of this residue at high temperature eliminates volatile substances, usually organic in nature, and decomposes bicarbonates with the loss of carbon dioxide.

The residue following ignition therefore contains the total inorganic solids, the difference between it and the original residue being termed the *loss on ignition*. The amount of residue in each operation and the loss on ignition is expressed as the proportion of the original water sample in terms of parts per thousand (ppt) or parts per million (ppm) depending upon the concentration, or as mass in metric units per liter (mg/l).

The *salinity* of fresh waters is defined as the total concentration of the ionic components. Although common in marine sciences, this term has not been widely used in freshwater researches. Because of the relatively small quantities of total ions generally encountered, freshwater salinity is usually expressed in milligrams per liter.

A measure of the total amount of ionized materials in water can be obtained through determination of the electrical conductance of the solution. Commonly called *specific conductance,* this parameter closely approximates the residue in solution and may be correlated with salinity. By definition conductance is the reciprocal of the resistance measured between two electrodes separated by 1 cm and having a cross section of 1 sq cm. The conductivity is usually expressed as micromhos (reciprocal megohms) per centimeter at 25°C. In general, the range of specific conductance of natural waters should approximate that of total dissolved solids given below.

In Hot Lake, Washington, at 3 m depth in August, the total dissolved solids amounts to around 391,800 ppm, and the specific conductance reads 60,440 micromhos. Most lakes occupying open basins have a total dissolved solid concentration of between 100 and 200 ppm. Evaporation from lakes in closed basins raises the concentration of dissolved solids, in some cases, to over 100,000 ppm. These lakes with extremely high concentrations are considerably more saline than the sea, the latter averaging about 35,000 ppm.

Organic compounds such as the organic states of phosphorus and nitrogen, sugars, acids, and vitamins are known in natural waters. We are very much lacking, however, in our knowledge of the formation and role of many of the organic substances. Although we shall relate dissolved substances to community metabolism in more detail in later chapters, it is well to keep in mind the idea that the quantity and quality of dissolved solids often determine the variety and abundance of plants and animals in a given aquatic situation. In a most general sense the limiting nature of dissolved solids is essentially two-fold.

In the first place the chemical density of the environment of aquatic organisms is a function of the total dissolved solids. According to laws of osmosis and diffusion the water balance, or osmoregulation, of plants and animals is governed by this environmental density factor and the physiological adaptations of the organisms to it. As we shall see later in more detail, osmoregulation is a very salient process in limiting marine and freshwater organisms to their respective habitats.

The second way in which dissolved solids influence the nature of the community relates to supply of nutrients and otherwise important materials. The only source of nutritionally important ions available to phytoplankton is the reservoir of matter dissolved in the water. The nature of the animal community is, of course, dependent upon the kinds and quantity of phytoplankton.

In another sense certain animals may be directly limited by the availability of a given dissolved substance, animals bearing carbonaceous shells being an example. In highly acid waters some mollusks may be entirely absent, while under mildly acid conditions mollusks may occur but their shells are reduced in thickness as compared with those inhabiting waters of higher pH value.

With these general concepts of the importance of dissolved solids in community function and structure in mind, let us now consider some of the properties and activities of the more abundant and better known inorganic and organic substances in natural waters.

SOLIDS IN SOLUTION IN LAKE WATERS

CALCIUM AND MAGNESIUM

Together, these two alkaline earth metals constitute the most abundant ions in fresh waters. The chemical activity of the two elements is similar, particularly in the formation of carbonate salts, and both may limit biological processes in streams and lakes. Magnesium is an important component of the chlorophyll molecule. Of these two ions, calcium is usually more abundant. In soft waters, or those containing less than 50 ppm of dissolved solids, calcium makes up, on the average, about 48 per cent and magnesium about 14 per cent of the total cations present.

In average hard waters the proportion of magnesium to calcium increases, giving approximately 53 per cent of calcium and 34 per cent of magnesium. The increase of these ions in hard waters takes place at the expense of two alkali metals, sodium and potassium, to be considered later. The nature of the lake basin is reflected in the ionic proportions of the waters in open lakes (those with considerable inflow and outflow). In this type of lake, calcium averages about 63 per cent; magnesium, on the other hand, makes up slightly more than 17 per cent of the total cations.

The wide variation in calcium content and the corresponding correlations shown in hardness, stratification, and biological productivity have stimulated many attempts at classification based on this ion. The following scheme was proposed by a German limnologist, W. Ohle:

< 10 mg Ca/l	"Poor"
10 to 25 mg Ca/l	"Medium"
> 25 mg Ca/l	"Rich"

Since calcium normally occurs in combination with the carbonate anion, the above classification should correspond generally with the designations of soft and hard waters given previously. Being concerned with the ionic content, both sets of designations are equally applicable to streams and lakes.

The distribution of calcium in stratified lakes generally follows a more or less characteristic pattern. In soft-water lakes of northern Wisconsin, the calcium content ranges from 0.7 to 2.3 mg/l (poor) and during summer stagnation little stratification of calcium exists. Lakes of medium hardness and medium calcium content typically contain a moderately increased calcium concentration in the hypolimnion during stratification. Such has been shown for lakes in southern Wisconsin which ranged from 21.2 to 22.4 mg Ca/l.

The hypolimnion of hard-water (rich) lakes characteristically contains a greatly increased load of calcium. In some of these lakes the curve of calcium concentration shows a steady increase of the ion from the surface to the upper region of the hypolimnion, the hypolimnial content being on the order of twice that at the surface and rather uniform throughout the stratum.

Under other conditions the epilimnial calcium may be relatively uniform throughout, increasing rapidly with depth through the metalimnion, and with less rapid, but none-the-less increasing change with depth in the hypolimnion. Along with other lake substances calcium is redistributed throughout the lake during vernal and autumnal mixing. An exception to calcium stratification in calcium-rich lakes is seen in data from Cultus Lake, British Columbia. This lake contains 32 mg Ca/l during summer but shows a somewhat uniform vertical distribution of the ion during stratification.

Biologically, soft-water lakes usually contain less living matter per unit area than hard-water lakes. The total estimated biomass of plant substance in certain medium lakes of Wisconsin was found to be three to five times that of poor lakes, while the animal mass, not counting fishes, was as much as triple that of poor lakes. Although the total mass of organisms may be greatest in rich lakes, medium lakes often harbor a greater variety of kinds of plants and animals. In addition to the quantitatively meager mass of organisms, poor lakes generally support a unique assemblage of species.

CALCIUM CARBONATE

We have already seen the interrelationships involving the carbon dioxide-bicarbonate-carbonate equilibrium, water, and the hydrogen ion concentration. If the carbonate formed by dissociation of carbonic acid combines with calcium (or magnesium) ions in the water we arrive at calcium carbonate as follows:

$$CO_3^{2-} + Ca^{2+} \rightarrow CaCO_3$$

Calcium carbonate, or lime, occurs as a white precipitate. Its formation is dependent upon the loss of carbon dioxide from the carbon dioxidecarbonate system. Recall that the addition of carbon dioxide serves to produce carbonic acid which dissociates and lowers the pH (increases the hydrogen ion concentration), thereby leading to acid conditions. The effect of carbon dioxide removal from the system is to split off CO_2 from HCO_3^-, which brings about the precipitation of $CaCO_3$.

In fresh waters calcium carbonate deposits, generally, are called marl, although some may be only accretions of animal shells. More properly, perhaps, marl is the accumulation of calcium carbonate precipitated directly from the water. This precipitation results from the removal of carbon dioxide from the water by physical loss to the atmosphere and by photosynthesis in green plants.

It appears well established now that many species of algae and some of the so-called higher plants are capable of precipitating lime in the process of obtaining carbon dioxide for photosynthesis. In some of these plants the carbon dioxide is taken directly from bicarbonate after the free carbon dioxide content has been exhausted.

Among the higher plants, *Elodea (Anacharis)* and *Potamogeton* often possess dense coatings of lime on their surfaces.

In fact, it has been found that 100 kg of fresh *Elodea* can precipitate 2 kg of $CaCO_3$ in a day with 10 hr of sunlight. Green algae (*Chlorophyta*) such as *Chara* and *Cladophora* under hard-water conditions cause the formation of considerable quantities of marl. *Chara*, or stonewort, feels brittle to the touch and some 30 per cent of its dry weight is lime. *Cladophora* often occurs in lime balls littering the lake bottom. Blue-green algae (*Cyanophyta*) and bacteria are also capable of forming lime.

Plant activity may not be the only process in the formation of lime deposits in fresh waters. It has been shown that carbon dioxide loss is also related to temperature, hydrogen ion concentration, partial pressure gradient across the water surface, and the nature of other substances in the water. Recently, it has been found that phosphate deficiency may be important in reducing the carbon dioxide content, thereby facilitating lime precipitation.

Calcium and magnesium enter into combination with anions other than carbonate, some of these being found in high concentrations in certain lakes. In the Swiss Alps several small lakes are rich in calcium sulfate ($CaSO_4$), or gypsum, derived from local deposits, Magnesium also combines with sulfate to form epsomite ($MgSO_4 \cdot 7H_2O$), which occurs in high concentration in certain saline lakes. The bottom of Hot Lake, a meromictic lake in Washington, is underlain by a laver of epsomite 4.5 m thick, from which epsom salts have been mined. A layer of gypsum is also found, lying below the epsomite stratum.

Unlike the "typical" water we have been emphasizing, this lake lacks carbonates, but the concentration of sulfate in the deep zone (monimolimnion) is over 243,000 mg/l. The same layer contains slightly over 700 mg Ca/l, and nearly 54,000 mg Mg/l. In other types of saline lakes magnesium may occur in combination with chloride as $MgCl_2$. Dolomite, a double carbonate of calcium and magnesium ($CaMg(CO_3)_2$), serves as an important source of both cations in waters in various parts of the world.

THE CALCIUM CYCLE

From our study of lime relationships it is obvious that calcium and carbonates are derived almost entirely from sedimentary rock strata. The ions are dissolved out of these formations and carried in solution in lakes and streams. Biological activity such as shell construction, bone building, and plant precipitation of lime combines and concentrates the ions. Through streams, much of the calcium reaches the sea, where it becomes "locked up" in coral reefs and bottom deposits of animal shells.

Here the calcium remains until geologic forces raise the sea-covered deposits and present them to the attacking forces of erosion and solution. The famous white cliffs of Dover, England, and the rich limestone regions of the Alps are examples of now-available sources of calcium in the perpetuation of the cycle.

SODIUM AND POTASSIUM

Where sodium and potassium occur in low concentrations the proportion of sodium is usually only slightly greater than that of potassium. As the total content of both increases, the concentration of sodium greatly, exceeds that of potassium. In average soft waters the equivalent percentage of sodium is second to that of calcium. In hard waters the proportion is less, usually falling below calcium and magnesium.

The most common form of sodium in natural waters is as halide (NaCl), reaching high concentrations in saline lakes such as Great Salt Lake with its salt content ranging from 24 to 26 per cent. Certain lakes in California, and in other regions of the world, contain sodium in the form of dissolved sodium tetraborate ($Na_2B_4O_7$), or borax.

Sodium sulfate (Na_2SO_4) occurs in abundance in some Canadian lakes. In certain of the saline lakes in Nebraska, potassium accounts for about 23 per cent of the total ionic composition.

Because of analytical difficulties, sodium is frequently determined and reported with potassium. It appears that the limnological importance and behaviour of these elements are similar.

NITROGEN AND NITROGEN COMPOUNDS

The very high importance of nitrogen in aquatic ecosystems rests upon its role in the synthesis and maintenance of protein, which is, along with carbohydrates and fats, a major constituent of living substance. Derived originally from the atmosphere, nitrogen enters into a complex cycle involving plants and animals and several forms of the element.

Nitrogenous compounds in natural waters may be derived from outside sources or may be fixed within the body of water (autochthonous).

The former category includes precipitation falling upon the earth carrying its own compounds (such as nitrate and ammonia), surface runoff containing terrestrial compounds of nitrogen (including pollution by human agencies), and inflow of ground waters as springs or seepages.

Endogenous nitrogen compounds result from fixation processes carried on by certain bacteria and algae. However, the extent to which "in place" biologically fixed nitrogen contributes to the total supply is not yet known.

In other words, we have yet to determine how much of the nitrogen content of a given lake is fixed within the lake by its own organisms and how much is delivered to the lake from the outside.

Once in the system, a great proportion of these compounds is caught up in a cycle of biological assimilation and decomposition, and inorganic processes in the economy of the ecosystem. Depending upon the efficiency of the community, certain quantities of these and other compounds will be lost. Lake, stream, and estuarine communities can lose nutrient materials through outflow, by incorporation of the substances in sediments, and through processes of denitrification releasing elemental nitrogen.

There are at least two possible sources of elemental, or uncombined nitrogen (N_2) in natural waters. One reservoir, and very likely the more important, is the atmosphere. The second, but poorly known, source of uncombined nitrogen is that produced by bacterial denitrification of ammonia. This uncombined nitrogen is rather inert; the only organisms capable of using it are certain microorganisms such as the nitrogen-fixing bacteria and algae.

The solubility of molecular nitrogen in fresh waters is related to temperature and pressure, the temperature relationship being an inverse one.

Although little study has been made of nitrogen in lakes, it has been established that supersaturation can exist under certain pressure conditions at the air-water interface. During summer stratification, the vertical distribution of uncombined nitrogen follows inversely that of temperature. Mixing, during vernal and autumnal overturn, distributes the nitrogen throughout the lake.

The synthesis of inorganic substances into plant and animal tissues and the metabolic processes of protoplasm produce various compounds containing nitrogen. These *organic nitrogen compounds* include, for example, nitrogen in combination with carbon and other elements, animal and plant protein, and urea and uric acid as animal metabolic wastes. Of the total content of soluble nitrogen in filtered and centrifuged surface waters of lakes, it is probable that 50 per cent or more is in the form of organic nitrogen.

Some 60 to 80 per cent of this organic nitrogen is composed of amino compounds such as free amino acids, polypeptides, and proteins. These substances are contained primarily in living plants and animals, and the presence of the compounds in water doubtless reflects metabolic processes of living organisms as well as decomposition of dead bodies.

With respect to formation by living organisms, it has been established that many blue-green algae secrete extracellular nitrogenous materials, including polypeptides, amides, and amino acids. The ecological importance of such liberated compounds is not fully known. In the algal species tested, the plants were not able to utilize their own excreted products as nitrogen sources.

The concentration of organic nitrogen may be expected to vary seasonally. Figure, shows the seasonal distribution of various forms of nitrogen in Lake Mendota, Wisconsin. There is little evidence that much of the organic nitrogen is available as nutrient for plants and animals. The measure of the total organic nitrogen content is, however, a valuable indication of the productivity of the body of water, for certainly most of the substance will ultimately be transformed into states which can enter into production of living matter.

In addition to its occurrence in the uncombined state and in organic compounds, nitrogen is also present in natural waters in the form of *inorganic nitrogen compounds* such as ammonia, nitrite, and nitrate. In most fresh waters the concentrations of these inorganic compounds are relatively slight, but nevertheless very important in determining the productivity of a given community. All of the inorganic forms can be used by most green plants, particularly various algae in their role of primary producers of energy-containing mass that can enter the aquatic food web.

Nitrogen "locked up," so to speak, in organic compounds is returned to the environment through decomposition and, to a lesser extent, by excretion of nitrogenous wastes of animals. The end product of the first stage of oxidative degradation of animal and plant proteins is mainly free ammonia (NH_3). Lesser amounts of ammonium compounds such as the base, ammonium hydroxide, and a salt, ammonium carbonate, are also released. The agents of the decomposition process are microbial organisms, i.e., certain bacteria and fungi. The free ammonia content of natural waters is derived in part from this bacterial decomposition of proteins, and in part from deamination (the removal of an amino group (NH_2) from an amino acid) also involving bacteria.

Generally, in unpolluted waters ammonia and ammonium compounds occur in relatively small quantities, usually on the order of 1 mg/l or less. With the uptake of oxygen, as in pollution, however, the concentration of ammonia may increase, in extreme cases to 12 mg/l or more. The relationships between ammonia and its basic compound, undissociated ammonium hydroxide (NH_4OH), seem to rest upon the pH of the water. At a temperature of 18°C and pH 6 the proportion of NH_4^+ to NH_4OH is approximately 3000:1, while at pH 8 the proportion is nearer to 30:1.

Under highly alkaline conditions the concentration of NH_4OH may reach toxic levels. Free ammonia in concentrations over 2.5 mg/l in neutral or alkaline waters is apt to be harmful to many freshwater species. The common ammonium salt of fresh waters, ammonium carbonate, is usually present in small amounts, but in concentrations of about 20 mg/l or more under alkaline conditions it is also toxic to certain animals.

Although somewhat variable, the ammonia content of the upper, lighted zones of unpolluted lakes is generally low, particularly in stratified lakes in summer. During summer stagnation the hypolimnion of eutrophic lakes may become considerably enriched with ammonia from decomposition in the tropholytic zone. With the advent of autumnal circulation the hypolimnetic ammonia is distributed throughout the lake, thereby increasing the content in the upper strata.

Similarly, ammonia produced during winter stagnation is circulated throughout the lake during vernal overturn, producing another seasonally high concentration of the compound. Spring and summer growth of phytoplankton populations in the trophogenie zone, and the resultant increased biochemical demand, bring about a reduction of ammonia during summer and up to the time of fall overturn.

In an intermediate phase of oxidation of nitrogenous organic compounds, ammonia is attacked by certain of the nitrifying bacteria, such as *Nitrosomonas*, which absorb ammonia and release nitrite (NO_2^-) ions in the process. Energy is released in this activity and is used in the synthesis of carbohydrates by the bacteria. Nitrite may also be formed by reduction of nitrate (NO_3^-), and it is probable that more of the nitrite content of most natural waters is derived from this process than from oxidation of ammonia.

Nitrite nitrogen, as it is usually designated, occurs in very minute quantities (if at all) in unpolluted waters. In lakes in which nitrite nitrogen is present, seasonal variation in concentration of the compound appears to follow (except in early winter) that of nitrate from which, as we have seen, the nitrite is probably formed.

In Lake Mendota the lowest content was found during late summer, when nitrate was also at a minimum. It is now known that diatoms and certain algae, *Chlorella* for example, are capable of reducing nitrate to nitrite. Since all green plants require nitrate, the supply of this compound may become quite low toward the end of the growing season.

It seems reasonable to assume, therefore, that nitrate reduction to nitrite would decline during the same period. During early winter, when nitrate and ammonia concentrations are low, nitrite content may increase; the reason for this is not known.

Vertically, the concentration of nitrite is usually maximum between the oxygen-rich trophogenic region with high nitrate content toward its lower boundary, and the oxygen-poor tropholytic zone often high in ammonia. The ecological importance (if any) of small nitrite quantities is not fully understood. Large amounts of the compound usually indicate pollution by sewage.

In what might be considered the final phase in the ecologically important process of decomposition of nitrogenous substance, nitrite is oxidized by nitrifying bacteria to nitrate (NO_3), expressed usually as nitrate nitrogen. It is in this form that nitrogen is most easily taken up by green plants rooted in the substrate or floating in the water. To the bacteria, *Nitrobacter* for example, involved in the process, formation of nitrate is of less importance than the energy released for their own utilization. To the community and its maintenance and productivity, however, nitrate is extremely important as a nutrient in supplying nitrogen for protein synthesis.

Nitrate nitrogen usually occurs in relatively small concentrations in unpolluted fresh waters, the world average being 0.30 ppm. In certain highly saline lakes (those of magnesium sulfate, for example) this form of nitrogen may be entirely absent at times. Lake Superior has a total dissolved solids content of about 60 ppm, making it, in this respect, the lowest of the Great Lakes. Nitrates account for 0.86 per cent of the total in the lake, and this represents the highest concentration of all the Great Lakes. Under the influence of edaphic factors, particularly during flood times, and organic pollution, nitrate nitrogen content may be expected to increase significantly.

Under normal conditions, however, the amount of nitrate in solution at a given time is determined by metabolic processes in the body of water, i.e., production and decomposition of organic matter. As we have seen, nitrate is contributed to the ecosystem as a byproduct of bacterial nitrification. The compound is removed from solution through utilization by green plants, and through bacterial denitrification to uncombined nitrogen and reduction to ammonia nitrogen.

Seasonal fluctuations in nitrate nitrogen and other forms in Lake Mendota, Wisconsin, are shown in Figure. Annual cycles in lakes, generally, may be expected to vary as to time of maximum nitrogen production depending upon latitude, basin morphology and chemical nature of the drainage area substrate, and productivity. Vertical distribution of nitrate is apparently related to lake productivity.

In some oligotrophic lakes that have been studied there is little evidence of nitrate stratification; in others the nitrate content of the upper zone decreases during summer from the amount present at spring overturn, with little change in deep waters. In eutrophic lakes, the nitrate concentration is typically decreased in the upper zones by plankton utilization and in the deepest regions by bacterial reduction; the result is a high content near the lower limits of the trophogenic zone.

PHOSPHORUS

In ecological thinking, phosphorus is often considered the most critical single factor in the maintenance of biogeochemical cycles. This extreme importance stems from the fact that phosphorus is vitally necessary in the operation of energy transfer systems of the cell, and that it normally occurs in very small amounts. The latter factor means that there is apt to be a deficiency of the nutrient, and this in turn could lead to inhibition of phytoplankton increase, resulting ultimately in decreased productivity in the system.

Phosphorus is known to occur in several forms, those of greater concern in natural waters being soluble phosphate phosphorus, soluble organic phosphorus, and particulate organic phosphorus of the seston. In water, phosphate may enter into combination with a number of ions but more conspicuously perhaps with iron and with the usually abundant calcium.

The pH of the water determines to a great extent the nature of the phosphate compound. Under circumneutral and moderately alkaline conditions calcium phosphate is probably prevalent, while extremely high pH usually results in the formation of sodium phosphate. In acid waters phosphate attraction swings toward iron to form ferric phosphate. Except for precise investigations of the activity of the ion itself or of biological assimilation of the various fractions, it usually suffices in general limnological studies to report total phosphorus, or phosphate.

The concentration of total exchangeable phosphorus in natural waters is determined primarily by:

- Basin morphometry as it relates to volume and dilution, and to stratification or water movements;

- Chemical composition of the geological formations of the area as they contribute dissolved phosphate;
- Drainage area features in relation to introduction of organic matter;
- Organic metabolism within the body of water, and the rate at which phosphorus is lost to sediments.

Ground waters and flowing surface waters are typically richer in inorganic phosphate than are surface waters of open lakes, due mainly to less biological demand in proportion to water volume.

In most open lakes assimilation by phytoplankton and bacteria serves to reduce the inorganic phosphate content. In closed basins of arid regions evaporation may result in very high concentrations of total phosphate.

With respect to geological influence, it should suffice to say that waters in local regions of highly phosphatic substrate contain considerable quantities of the ion. It is doubtless safe to state that all bodies of water that support some plant populations contain a quantity of phosphate, albeit small, in some cases less than 0.001 ppm (frequently expressed as milligrams or micrograms per cubic meter because of the minute quantities normally encountered).

The seasonal distribution of phosphorus in lakes is variable, being determined to a great degree by basin form, chemical composition of the surrounding terrain, land use, behaviour of other substances in the particular lake, and the annual cycles of mixing.

In relatively open basins receiving considerable influx of surface water, phosphorus concentration is often regulated by stream discharge, particularly following high rainfall. The Maumee River drainage of Ohio annually contributes on the order of 96 metric tons of soluble phosphate to the western basin of Lake Erie. Of this total, 68 per cent is carried in seasonally during January and February, the time of greatest stream discharge.

The Maumee load is apparently derived as leaching from the highly agricultural region, for other streams of the era transport proportionately less phosphorus. In lakes of more self-contained dynamics seasonal variations in total content and vertical distribution of phosphorus are rather completely related to other compounds and mixing cycles.

In eutrophic lakes during summer stratification the phosphorus content in the hypolimnion increases significantly following oxygen depletion. The factors involved in this increase are not fully understood. One investigator has shown that this phosphorus is apparently released from a ferric iron-phosphorus complex which is insoluble in the presence of oxygen; in the absence of oxygen the ferric compound is reduced to a soluble ferrous form, thus liberating phosphorus.

With lake overturn and the reintroduction of oxygen the insoluble ferric phosphate is again formed and distributed throughout the lake. From another source has come evidence that in the presence of oxygen phosphorus is adsorbed on basic iron compounds in the oxidized microzone of the mud. Removal of oxygen brings about a reaction in which the ferric ion is reduced and phosphorus is released.

While it is possible that both mechanisms operate, it seems certain that the presence or absence of oxygen is a critical factor. Soluble inorganic phosphorus in the upper waters of eutrophic lakes is usually low throughout the year, becoming depleted at times during summer. The concentration of total phosphorus (mainly the organic form) may increase in late summer.

THE CYCLE OF RADIOACTIVE PHOSPHORUS IN LAKES

We are all aware of the astounding array of new tools of everyday living and instruments of research made possible by the twentiethcentury exploitation of atomic energy. Some of the materials and techniques have already entered into limnological investigations, and many more are destined for future usefulness. Prior to the "radioactive era" our knowledge of phosphorus activity in natural waters was based on field and laboratory analyses and measurements of seasonal variations in concentration of the several forms of phosphorus, and the distribution of the substances in natural waters.

Now through the use of radioactive phosphorus (P^{32}) we know quite a bit about the rate and processes of phosphorus circulation in at least certain types of natural waters. Even so, our knowledge at this stage is still meager, and much is vet to be learned. Actually, radioactive phosphorus is only one of many such materials presently available for use in new and exciting research in the dynamics of natural communities, be they aquatic or terrestrial. Recently, isotopes of calcium (Ca^{45}), strontium (Sr^{90}), and yttrium (Y^{90}) have been utilized in researches bearing on biological limnology.

Of the many questions asked in investigations of nutrient cycles, those of utmost importance concern the source of the nutrient, the forms in which the nutrient occurs throughout the cycle, the manner of transformation of the fractions, and the rate at which the nutrient is circulated. In the case of phosphorus, for example, the major forms such as we have just considered have been known for some time. How are these forms produced and at what rate?

We also know that when phosphorus fertilizer is dumped into a lake most of the compound disappears rapidly, the disappearance sometimes being acompanied by increased plant production and sometimes not. How do we account for this? Among other things, we are aware of the mud-water differential in phosphorus concentration, but as indicated previously we are not sure of the causes, or the rate of turnover of phosphorus from these and other sources.

If a quantity of radioactive phosphorus (P^{32}) is diluted to a desired strength of radioactivity (usually as counts per minute) and added to a body of water or used in laboratory experiments, the isotope enters into the system under study. The fate of the introduced (P^{32}) can then be determined as radioactivity of inorganic and organic phosphorus in living and nonliving components of water and mud. The rate at which the isotope is transferred within the system may be measured also. Let us now review some of the findings from experiments designed along those lines.

In the first place, it was found that radiophosphorus, when added to a lake surface, decreased at a high rate as had been observed for fertilizer. Within a few days between 70 and 90 per cent of the (P^{32}) was lost from the epilimnion.

This led to the idea that phosphorus in lakes is held in a kind of repository from which the nutrient is withdrawn to maintain a steadystate equilibrium.

In other words, there probably exists a sort of reciprocal exchange relationship between phosphorus in the water and its contents, and phosphorus in solids in contact with the epilimnion (recall that during stratification there can be little mixing of bottom-produced substance in the epilimnion).

This state of equilibrium leads to the question of rate of exchange, or turnover time, between the two phases. Turnover time, in this case, refers to the rate at which atoms of phosphorus move from one phase to another, as for example from the lake water to the bottom mud.

More precisely, it is the time during which as many atoms are transferred through a phase as there are atoms in that particular phase. The use of tracers permits recognition of atoms of a given phase.

A cycle of phosphorus, including turnover times of the components, is shown in Figure. The exchange rate between mud surface and water, at the bottom of the figure, was determined from laboratory experiments rather than from studies of a lake—in which the rates are apt to be rather variable. Exchange rates in the water phase are probably constant within lakes generally. Of particular interest is the rate at which the introduced inorganic (P^{32}) is taken up.

Note that there is considerable competition among several factors for inorganic phosphate. Laboratory experiments indicate that over 95 per cent of the original radiophosphorus may be taken up within 20 min by phytoplankton and bacteria, the turnover time of the inorganic fraction in the epilimnion being on the order of 5 min.

Certain phytoplankton can convert inorganic PO_4 to the organic state in less than 1 mill. As shown at the lower right (and this is probably the same substance as at lower left) about 4 per cent of the planktonic cells that take up inorganic PO_4 are sedimented to the bottom each day.

Inorganic reactions involving the uptake of inorganic phosphate through the water-mud interface are given at lower centre, these reactions proceeding when bacterial activity is inhibited. It is here seen that the rate of return of the phosphate to the water (15 days) is much slower than the turnover from the water (3 days).

The activity of bacteria in phosphorus utilization revealed in these studies is of great ecological significance. It was found, for example, that these microorganisms take up large quantities of inorganic phosphate either by assimilation into their own bodies or by conversion to the organic fraction, thus making the nutrient unavailable for use by green plants. In the absence of bacteria, the rate of uptake of (P^{32}) by algae and rooted plants is very high. It appears that in the competition bacteria get a large share of the available inorganic phosphorus.

This in turn could seriously limit production of animal mass in a community, although we do not know to what extent these autotrophic bacteria are utilized by consumer organisms in the food web.

The problem of bacteria versus phytoplankton is difficult to assess, however, for we know very little about the phosphorus requirements in the maintenance of natural populations of the various algal species. Many studies have been made of this subject of plant food requirements, but in most cases laboratorydetermined minimal concentrations necessary for growth have been higher than the average content of most natural waters. On the other hand, it has been shown that an excess of phosphorus inhibits plankton growth in laboratory cultures.

IRON

Iron is of particular interest because of its importance as a vital element in respiratory pigments of many animals, and because of its part in many chemical reactions in water. We have already seen various aspects of iron in relation to oxidation-reduction systems, and its reactions with carbonate and phosphate compounds.

Whether or not these reactions enter into prominence in particular bodies of water depends upon the presence of the compounds other than iron and the momentary form in which the iron occurs. Iron is found widely in nature, usually as either bivalent Fe^{++} or trivalent Fe^{+++}. The bivalent, ferrous, state is soluble, but only under anaerobic conditions. In the presence of oxygen the trivalent, ferric, form is present as a colloidal complex in combination with other inorganic ions and simple decomposition products.

With oxygen depletion, as for example in the hypolimnion in summer, the ferric form is reduced to ferrous, the latter going into solution. As a result of the breakdown of the ferric complex the concentration of silicate, phosphate, bicarbonate, or iron is often increased, depending, of course, upon the original chemical nature of the waters.

It might be pointed out here that the mere absence of oxygen will not bring about the transformation from ferric to ferrous iron. The depletion of oxygen is a result of organic decomposition which also forms organic compounds that reduce the ferric iron.

In some instances, that of ferrous bicarbonate for example, high carbon dioxide content and near-neutral or acid conditions are necessary in addition to absence of oxygen and presence of reducing substances of organic origin.

The vertical distribution of iron in lakes is primarily a composite picture of several forms of the element influenced by the solubility factors just considered. In view of the chemical conditions prevailing in the epilimnetic regions of most lakes the iron content, maintained by several forms, is quite low, usually less than 0.2 ppm.

In the deeper zones the iron concentration is a function of oxygen content and redox potential. Where the hypolimnion is enriched with oxygen, as in oligotrophic lakes, the iron is normally contained as the nonsoluble ferric complex. If oxygen becomes deficient at the mud-water interface, ferrous iron may go into solution.

In eutrophic lakes the hypolimnion typically contains iron in solution during the later stages of summer stagnation. Even at this time the content is normally low, generally a few parts per million.

One of the forms of iron which appears to be most readily available to phytoplankton is ferric hydroxide ($Fe(OH)_3$). This form is the usual product of the oxidation of ferrous iron in waters containing dissolved ferrous salts. It normally occurs as an amorphous mass of a ferruginous organic complex. In the utilization of the organic component of the complex by bacteria, ferric hydroxide is precipitated as a by-product. This process also leads to the formation of lake ochre, the "parent" of iron ore.

In the filamentous bacteria of the genus *Leptothrix*, iron hydroxide is left in sheaths surrounding the microbe as it metabolizes the organic material. The species of *Gallionella* excrete ribbon-shaped strands of ferric hydroxide that attach to objects in the water. This and such forms as *Ochrobium*, *Siderocapsa*, and others probably derive energy from the oxidation of ferrous iron.

SULPHER, SILICA, AND OTHER ELEMENTS

SULPHER

The most frequently encountered forms of sulpher in fresh waters are as the anion sulfate (SO_4^{2-}) in combination with the common cations, and as hydrogen sulfide (H_2S). Sulfate enters bodies of water with rain, and through solution of sulfate compounds in sedimentary geologic formations in the drainage area.

With respect to the latter source we should add that sulfate is not highly soluble. Surface waters are generally low in sulfate, except in regions locally rich in the ion, and in closed lake basins where evaporation raises the concentration. In some geographic regions rain may be the major source of the ion. The average percentage content of sulfate in North American waters is about 15 per cent.

Although this quantity places the ion second in abundance, it is also less than half that of carbonate. The low sulfate content of Bunny Lake is probably characteristic of youthful lakes (geologically speaking) formed in regions of weakly soluble terrain. Purcell's Pond is in an area of granitic rocks which normally contain little sulpher in proportion to tile alkali and alkaline earth metals.

However, this lake has a large peat bog along one side, and, since *Sphagnum* is known to accumulate sulpher from the atmosphere and release it as sulfuric acid, this contribution would augment that from rainfall. Nevertheless, the content of 9.5 mg S04/l for this pond is quite small compared with that of Hot Lake, Washington, which amounts to 103,680 mg S04/l at the surface and 243,552 mg/l at the bottom.

Sulfate is ecologically important in natural waters in several ways. It is apparently necessary for plant growth; short supply of the material can inhibit the development of phytoplankton populations and, therefore, production. Sulpher is important in protein metabolism and is supplied to the organisms originally as sulfate.

As we shall presently see, under anaerobic conditions sulfate is utilized in chemosynthetic processes of sulpher bacteria, the sulfate being reduced to hydrogen sulfide. This in turn may result in the liberation of phosphates and other nutrients held in ferric complexes.

Seasonal pulses in sulfate concentration have been demonstrated in certain lakes. In some, the sulfate reaches a high in spring followed by decreasing content to a low in autumn. Bicarbonate variation in the same pond behaves in the reverse fashion. It has been suggested that the sulfate decrease is due to reduction to sulfide which becomes taken up in the bottom mud; the bicarbonate fluctuation apparently results partly from carbon dioxide removal by photosynthesis and partly from sulfuric acid activity in the winter oxidation of sulfide.

In relation to the sulpher cycle, however, we should here note the activity of bacteria in maintenance of the cycle. The reactions involving H_2S are essentially between oxidation and reduction. Hydrogen sulfide can be oxidized by a number of so-called "colorless sulpher bacteria," among which *Beggiatoa* is probably the best known. The species of this genus are large filamentous forms often said to be more closely related to blue-green algae than to bacteria. Under normal conditions the cells of the filaments include granules of elemental sulpher accumulated only when both oxygen and H_2S are present. When H_2S becomes exhausted from the environment, the sulpher in the granules is metabolized by the organism and sulfuric acid is released.

The cellular oxidation exhibited by these bacteria apparently proceeds as follows:

$$2H_2S + O_2 \rightarrow 2S + 2H_2O + \text{energy}$$

$$2S + 2H_2O + 3O_2 \rightarrow 2SO_4^{2-} + 4H^+ + \text{energy}$$

Another group of bacteria, particularly the genus *Thiobacillus*, is obligately chemotrophic and can oxidize elemental sulpher to release sulfuric acid. Certain other bacteria, including the euphoniously named *Desulfovibrio desulphuricans*, can reduce sulfate (and other sulpher compounds), resulting in the release of H_2S. With this activity the cycle of consumption and production of H_2S is completed.

SILICA

Over 60 per cent of the rocks and soils at the earth's crust is silicon dioxide (SiO_2). In view of this proportion we should expect to find the compound in most unpolluted waters. As is the case with many other substances, silica is most abundant in sedimentary rocks and therefore generally occurs in higher concentrations in waters in regions of such rocks. Even within a given region, however, the silica content of waters may be quite variable. For example, in Florida, an area in which sedimentary soils and rocks predominate, the concentration of SiO_2 ranges from less than 1 mg/l to over 30 mg/l.

Unlike many other minerals, silica does not appear to be important in the composition of animal or plant protoplasm. It does occur, however, in the "shells," or frustules, of certain algae (the diatoms), in cysts of yellow-brown algae, and in skeletal spicules of some sponges.

Population development of such diatoms as *Asterionella, Melosira, and Tabellaria* is limited, at least partially, by concentrations of from 0.5 to 0.8 mg SiO_2/l. The utilization of silica by one or more of these organisms takes up the mineral from lakes and streams, and often during blooms of diatoms the quantity removed may be great.

OTHER ELEMENTS

In addition to those elements previously considered, natural waters may contain others in various proportions. Manganese is commonly found, and though less abundant, its role is similar to that of iron. Aluminum, zinc, and copper are usually found in natural waters in varying quantities. Traces of molybdenum, gallium, and nickel have been detected on occasion. Uranium appears to be common in small amounts in most lakes and streams, and man may be a strong environmental factor influencing future concentrations of this element.

ORGANIC SUBSTANCES

Organic matter is present in solution in water, and as fractions of plankton and organic debris, the seston, is suspended in water. Organic substances are also found, often abundantly, in bottom sediments. The total content of these materials in lakes is derived essentially from two sources:

- Income water and wave action at tile shore, which introduce allochthonous matter into the lake to account for a portion of the total—the nature of the substances brought into the lake from the outside being quite varied, ranging from leaf litter of the shoreline to upland soil leachings and organic pollution;
- Autochthonous matter produced in the lake by living organisms and by decomposition of plant and animal bodies.

The relative contribution of each of these sources to the lake total is of considerable importance in maintaining productivity. It would, therefore, be of interest to measure each, but at present we do not have precise, readily usable methods for doing so. Hutchinson has reviewed and commented upon several techniques. Data on the organic composition of a large number of lakes in Wisconsin have shown that the total content and the proportions of various fractions of the total organic concentration vary widely in lakes. In organically poor lakes the carbon content ranges from 1 to 2 mg per liter, and, of the total organic content, carbohydrates comprise slightly over 73 per cent and proteins about 24 per cent.

The richer lakes contain up to about 26 mg carbon per liter, with nearly 90 per cent of the organic matter being composed of carbohydrates and about 10 per cent of proteins. It appears that the protein content essentially decreases with increasing total organic content. In lakes poor in organic matter (little autochthonous substance) the seston organic material averages about 16 per cent of the total. In richer lakes seston matter comprises about 4 per cent of the total organic content.

The data from these studies indicate that the particulate organic matter of lakes increases with increase in dissolved organic substance up to a point; beyond this point the seston content does not seem to be influenced by additional dissolved matter. This relationship would seemingly point to the fact that autochthonous organic materials do not contribute to the productivity of lakes. In fact, it has been shown that the protein content generally is indeed low in the allochthonous portion of lake substances.

Lake colour may be correlated as a rule with the source of organic matter. Allochthonous material derived from peat or shallow-water plant debris is mainly responsible for the dark-brown colour of bogs. In such waters the protein fraction of the total organic content is low in proportion to carbon. In the waters of lakes receiving little allochthonous material the protein component is high in proportion to carbon, the autochthonous matter being formed mainly by plankton decay.

These waters are relatively uncolored. Colour and source of organic matter are apparently associated with plankton production also. In highly stained waters which contain considerable quantities of allochthonous substances, plankton may comprise as little as 3 per cent of the total organic substance. In unstained lakes receiving little allochthonous materials, over one fourth of the organic matter may be plankton.

The total organic content of most "average" lakes is as much as three or four times that of open seas. Sea water generally contains about 2 mg carbon and about 0.2 mg organic nitrogen per liter. These differences are of interest, but probably bear little on the composition of estuaries. Most estuarine basins receive considerable quantities of allochthonous materials from stream inflow and drainage from surrounding marshes or other land features.

An impressive array of organic compounds has been isolated, identified, or otherwise recognized in fresh and marine waters, bottom sediments, and soils. At present we know very little about the role of most of the organic compounds in the over-all ecology of freshwater or marine communities.

There is sound evidence that certain bacteria can utilize dissolved organic substances as an energy source. Beyond this it has not been established that muulticellular animals make use of such materials, even though a considerable controversy over the matter has existed for sonic 50 years.

Fats, in the form of simpler molecules of fatty acids, are known from freshwater and marine seston and water. Acids and certain waxes have been recovered from sediments. Some attention has already been given to organic nitrogen in natural waters. Other compounds such as free amino acids, tryptophan, glycine, glutamic acid, tyrosine, and others, are known from water, seston, and sediments. These have been recovered upon hydrolysis.

Although studied but little, some of these amino substances may be important to organisms which are unable to synthesize certain of the acids, the organisms obtaining the items directly from the environment.

Hydrolysis of seston and sediment components reveals a number of sugars and sugar-like compounds, the best known being glucose and galactose.

In estuaries a compound having the qualities of a reducing sugar has been shown to influence the water-pumping rate of oysters.

Yellow carotenoid substances of fluorescent qualities have been extracted from shallow marine waters, and from fresh waters. The marine compound may not, however, be the same as yellow acids of lakes. Fresh and marine waters contain biologically active compounds such as vitamin B12. The seston of both environments has yielded also vitamin A and vitamin D. Thiamin (B1), biotin (H), and niacin (B6) are known from lakes, the last-mentioned being present in amounts up to 0.89 mg/m3. Enzymes are present in natural waters and probably enter into various biological reactions in the environment.

It is now known that some organisms, particularly algae, produce metabolites that serve in various ways to inhibit, stimulate, or generally regulate the development of natural populations of other algae. The chemical nature of the particular metabolites is not known, although the possibility of hormone-like substances and fatty acids has been suggested.

DISSOLVED SOLIDS IN STREAMS

The reactions and behaviour of the various ions and organic substances in streams are governed by chemical and biological processes and conditions such as we have considered in the preceding pages. The main differences between lake and stream characteristics with respect to dissolved solids usually pertain to relative concentration, composition, and longitudinal distribution of the substances.

Seasonal variation in rainfall and surface runoff, and the geochemical nature of the drainage basin strongly influence the composition of waters of small streams, thereby imparting considerable individuality to streams even within a restricted region.

Variability becomes the key word in any attempt to categorize small streams. This variability is seen not only when comparing individual streams; it is also observed along the gradient of a particular stream, being markedly influenced by edaphic factors, human endeavors, and channel morphology.

Waters of large streams, on the other hand, typically exhibit general uniformity of composition, so much so that a quantitative expression of average content becomes meaningful.

The mean composition, as percentage proportions, of the major cations of river water has been given as: calcium 63.5, magnesium 17.4, sodium 15.7, and potassium 3.4. Note that the order of prominence here differs slightly from that of North American waters, generally. In fact, the average composition of rivers is quite similar to that of open lakes, as might be expected.

The calcium content of surface waters varies with pH, substrate composition, and temperature. Acid swamp streams and streams on igneous rocks contain little or no calcium, but streams in regions of sedimentary geologic features often contain high proportions of the ion.

Springs issuing from limestone frequently contain on the order of 70 ppm of calcium. This concentration is similar to that of many large North American streams; throughout much of its length the Missouri River contains, on the average, about 60 ppm of calcium. Much of the ion in large streams is carried to the sea, where it enters into the calcium cycle.

As in lakes, there also appears to be a correlation in streams between quality and quantity of algae, and calcium carbonate concentration. Under proper conditions calcium in streams may be precipitated as the monocarbonate by lime-encrusting plants, and some is taken up by shellbuilding animals.

The magnesium content of flowing waters is also highly variable, ranging from trace quantities to nearly 500 ppm in warm mineral waters. Although calcium and magnesium react similarly, particularly in entering into carbonate combinations, their proportions in freshwater streams vary from about 2: 1 to 10: 1 or more, respectively.

Nitrogen compounds normally exhibit conspicuous seasonal fluctuations and pronounced variations along the gradient in small streams. The amount of nitrate nitrogen and nitrite nitrogen is influenced to a great degree by surface runoff and associated stream level and discharge. Nitrogen in ammonium compounds is released into streams through decomposition of organic debris. In unpolluted streams the concentration is small, normally less than 1 ppm. Pollution raises the concentration of ammonium compounds, and this, within limits, increases biological productivity. In excess, certain compounds of ammonium can be toxic to stream organisms.

High concentrations of nitrogen occur during times of winter and spring flooding, when plant populations are minimal. Interestingly, the floods that serve to bring in nitrogen may also scour algae from rocks in the stream bed, thus minimizing consumption during the temporary increase in nitrogen.

We might add that rapid decrease in nitrogen follows closely behind the flush of flood water. In streams in more stable watersheds, the nitrogen content may be lowered in spring and summer by plant utilization, with continued decrease to late summer. The major source of nitrogen compounds in many streams is pollution by drainage from agricultural areas and from sewage. These factors contribute large quantities of the substances regionally and have noticeable effects on stream communities.

The phosphate compounds of streams are derived from biological and chemical processes along the stream course. During summer the concentration of inorganic phosphates may increase somewhat, due to biological activity. Increased surface runoff contributes to the phosphorus content of streams by introducing allochthonous phosphorus-containing substances.

Although there is considerable longitudinal and seasonal fluctuation in content, there does not appear to be any great depletion of stream phosphorus such as we have noted in surface waters of lakes during the growing season. This is due partly to mixing and partly to the normally greater proportion of water to phytoplankton. In fact, as we shall see later, there is little, if any, true plankton in streams of fair velocity.

Iron occurs naturally in streams, although in relatively

small proportions. In unpolluted flowing waters this ion usually takes the ferric form because of continual aeration and the presence of oxygen. Where organic decomposition is great, oxygen depletion may result in the transformation to the ferrous state and the precipitation of iron hydroxide. Similarly, in stagnant stream pools, particularly following subsidence of flood conditions, iron bacteria may grow rapidly and form masses of ferruginous substance in the pools.

Sulpher, as sulfate or hydrogen sulfide, occurs in varying quantities depending upon the composition of the substrate, the area of the catchment basin (its relation to rainfall), and the course of the stream. Obviously, the importance of the substrate rests on the proportion and solubility of minerals present. The stream course determines to a great extent the occurrence of hydrogen sulfide, for, if the course permits slow flow through organically rich swamps or marshes, tile gas may be formed by decomposition or, as we have seen, by the reduction of sulfate.

A number of additional ions, including those discussed in the preceding section, are present in streams. Their abundance is usually associated with land forms and chemistry, and the ecological role of many of them is poorly known. The importance of climate, geochemistry, and physiography in stream composition has not been studied sufficiently. We can see some of the interrelationships in a few summarizing statements relating to certain of the major ions.

The waters of rivers whose headwaters lie in semiarid plains are generally high in sodium, sulfate, and chloride, but low in calcium and carbonate. Streams draining youthful granite mountains typically contain high concentrations of silica, and the total dissolved solid content is low. Valley streams in granitic mountain regions show an increase in total dissolved solids. Rivers draining extensive plains in humid, temperate regions are characterized by high sulfate and carbonate concentrations.

SALINITY

Numerous references have been made to salinity in

preceding sections. These references have shown the influence of this factor on various Sea Water arine parameters such as density and stratification, temperature, and dis Sea Water solved gases. Let us now give attention to a more precise definition and to consideration of salinity in its more or less true light, that of its chemical aspects.

The ions together with strontium, bromide, and boric acid, contribute over 99 per cent of the total salts in solution in fullstrength sea water. The weight in grams of these salts dissolved in 1 kg of sea water is termed the salinity.

In the process of drying sea water in order to weigh the salts, carbonates are decomposed to oxide, and bromine and some quantity of chlorine are liberated; these transformations and losses are taken into consideration in the definition. Salinity is usually expressed as parts per thousand (‰), although per cent (%) and grams per kilogram are also used. Since the salts of the metallic ions, the conservative elements, occur in very uniform proportions, it is possible to determine the salinity by measuring the chloride content, or chlorinity and applying it in the relationship:

$$\text{Salinity ‰} = 0.030 + 1.8050 \times \text{chlorinity ‰}$$

Chlorinity is rather easily and accurately measured by silver nitrate titration, using potassium chromate as an indicator. Salinity can be determined by electrical conductivity, and from measurements of density obtained by the use of hydrometers.

The salinity of open seas generally ranges between about 33 and 38‰, with the average being near 35‰. Since the average salinity. of soft fresh water is 0.0651‰ and of hard fresh water 0.30‰, it is apparent that high estuarine salinities are derived almost wholly from sea water, while the diluting effect of the influent serves to reduce the concentration of dissolved salts.

In general, the proportion of dissolved salts in estuaries resembles that of sea water, while the total concentration is variable along the axis of the estuary. In some instances, however, the concentration of inflowing fresh waters may be such as to modify the normal ionic relationships in estuaries.

Where this occurs the result is usually an increase in the ratios of carbonate and sulfate to chloride, and of calcium to sodium over those of average sea water.

The momentary salinity may be regarded as a function of the quantity and quality of inflowing and outflowing waters, rainfall, and evaporation. Since these factors may vary with season (in some instances rather drastically), the general structure of the estuary also shifts. Therefore, attempts to fit estuaries into schemes of classification are often difficult. We can, however, recognize that certain estuaries are typically, or on the average, more or less saline than others. For example, the average salinity of the estuary-like Laguna Madre of Texas nearly always exceeds that of sea water.

Chapter 5

Relationships of Organisms and Environment

POPULATIONS IN AQUATIC ENVIRONMENTS

A group of organisms occupying a given space at a particular moment in time is called a *population*. A population may consist of a single species and thus constitute a *species population*, or it may be composed of a number of species exhibiting similar ecological traits (food requirements, breeding sites, etc.) and regarded as a *mixed population*.

In addition to structure and time-space attributes, a population also exhibits a number of measurable characteristics such as birth rate (natality), death rate (mortality), density, and capacity for increase. The extent to which these traits are manifested in an environment is strongly influenced by properties of that environment.

THE PRINCIPLE OF LIMITING FACTORS

The very fact that we can recognize an environment-organism complex such as a lake community or an estuarine community attests to the premise that all organisms do not react uniformly to all combinations of physical, chemical, and biological features of the environment.

Through evolutionary processes, organisms have become variously adapted to certain sets of conditions. It follows, therefore, that successful development and maintenance of a population depend upon harmonious ecological balance between environmental conditions and tolerance of the organisms to variations in one or more of these conditions. This idea suggests that one or more factors can serve to limit the areal distribution, density, and other attributes of a population. A factor that so checks or exerts some restraining influence upon a population through incompatibility with species requirements or tolerance is said to be a *limiting factor*. You will note that the limiting factor principle rests essentially upon two basic concepts. One of these relates organism to environmental supply of materials needed for metabolism and growth. The second concept pertains to the tolerance which an organism exhibits toward environmental factors and conditions.

The aspect of environmental supply in the limiting factor principle includes, among other things, the provision of nutrients and other substances (respiratory gases, for example) necessary for growth of the organisms. If, out of a broad array of materials required for growth and development, one is absent, a given species will not be able to survive. On the other hand, if the nutrient is present but only in small amounts, the population of a particular species will be limited proportionately. These conditions were described in 1840 by Liebig in what has come to be called the *law of the minimum*. This principle recognizes that the development of a population is essentially regulated by the substance occurring in minimal quantity relative to the requirement of the population.

It is frequently difficult to determine precisely which of the factors of a complex is exerting a limiting effect. This is due largely to the fact that the utilization of a certain substance may be regulated by other materials, or by physical factors in the environment, or by physiological adjustments of the organism itself. It has been shown, for example, that the uptake of phosphorus by the alga *Nitzschia closterium* is influenced by the quantity of nitrate and phosphate in the environment; nitrate utilization, on the other hand, appears to be unaffected by the phosphate. Potassium, calcium, and magnesium generally regulate interactions in plant metabolism. The assimilation of nutrients by some algae is related to environmental temperature. Temperature and carbon dioxide tension may affect the rate of oxygen utilization in certain fishes. We have previously learned that the absence of free carbon dioxide in waters is not necessarily critical for certain plants which can assimilate carbon dioxide from bicarbonates.

Dissolved oxygen is necessary in respiration of most aquatic organisms. The minimal requirements of species vary and often limit the spatial distribution of certain forms in the community. The water issuing from the "boil" is oxygenless, but gains the gas with downstream flow. A small amount of hydrogen sulfide is present in the boil and decreases downstream; this may also be involved in the observed distribution.

The law of the minimum pertains primarily to the reaction of organisms to factors necessary for growth and metabolism. In the case of plants, these factors include the inorganic nutrients and solar energy for photosynthesis. In addition to satisfying their metabolism needs, organisms are also confronted with physical and chemical factors which essentially regulate the distribution and numbers and the extent to which the organisms are able to utilize growth factors. Such environmental features as salinity, temperature, and currents govern many populations. The extent to which these factors limit a population depends primarily upon the *tolerance* of the organisms to a single factor, or, in some instances, to a complex of interacting factors. These ideas are the essence of the *law of tolerance* stated in 1913 by V. E. Shelford of the University of Illinois.

Tolerance to features of the environment varies widely among aquatic organisms. A population may exhibit a wide range of tolerance toward one condition, and a narrow range toward another. All stages in the life history of an organism do not necessarily show similar ranges of tolerance. The range of tolerance toward a given factor may be modified by another factor. A wide range of distribution of a species is usually the result of the wide tolerance.

In describing the tolerance of an organism, the prefix *eury-* meaning wide, or *steno-*, meaning close or narrow, is added to a term for the particular feature. We have already become familiar with the terms *euryhaline* (wide salt tolerance) and *stenohaline* (narrow salt tolerance). To these, we might add *eurythermal* and *stenothermal*, pertaining to temperature. Let us now consider some examples of tolerance relationships, for such are of utmost importance in regulating the abundance and distribution of organisms.

Contractile vacuoles are characteristic of freshwater protozoans, but not of their marine relatives. In certain flatworms (*Gyratrix*, for example) the flamecell mechanism serving in osmoregulation and excretion is more complexly developed in freshwater forms than in estuarine ones; in marine flatworms, major parts of the system are absent. Freshwater crayfish possess long, well developed nephridial canals; in the estuarine and marine lobsters these structures are reduced and poorly developed.

The fluids of marine bony fishes are hypotonic to the environment, and therefore the animals tend to lose water to the medium. In contrast to freshwater inhabitants, marine fishes actively drink water, excrete a concentrated urine, and excrete excess salts (obtained by drinking sea water) against an environmental gradient through specialized cells in the gill region. In the estuarine environment, these marine forms must function in the manner of freshwater species. The young of certain species of marine fishes inhabit estuaries and possess glomerular kidneys; with migration to higher salinity as adults, the glomeruli degenerate. Highly euryhaline fishes, particularly those adapted to extreme migrations from fresh to salt water (the eel (*Anguilla*), for example), are usually capable of performing as either marine or freshwater fish.

We have dwelt on this subject of osmoregulation because it demonstrates some of the forms of adjustment often necessary in order for organisms to tolerate certain controlling factors in the environment. In many instances, however, the limit of tolerance of a given organism may be modified by other factors. It has been shown, for example, that the shrimp (*Crangon crangon*) and estuarine crab (*Carcinus maenas*) are able to tolerate lower salinities during summer seasons of warm water. In Japanese waters, the oyster (*Ostrea gigas*) exhibits a wider range of salinity tolerance at winter temperatures than during the warm season. The tendency for marine species to tolerate lower salinities at higher temperatures accounts, in part at least, for the generally greater number of species in estuaries of the tropical zones than in the cooler climes.

As in the case of salinity, temperature also acts as a controlling factor related to range of tolerance of species. In this sense, temperature serves to regulate growth and metabolic rates of various organisms, and often determines the time of reproduction. It is immediately apparent, therefore, that temperature is highly important in delimiting the rate of utilization of nutrients and light plants, and the tempo of food intake by animals, to satisfy metabolic demands. These relationships are contained in the "Q_{10} Rule," or Van't Hoff's principle, a chemical law applied to some physiological processes.

The principle holds that the rate at which biological processes proceed is increased nearly two-fold with each 10° rise in temperature (the rate is a linear function of temperature). Obviously, this principle operates only within the range of tolerance of a given species, and is further restricted by an optimal point within the over-all range. For example, if sufficient nutrients are available, an algal population may grow rapidly and "bloom" at the optimal temperature for that species; beyond this point, and before the maximum tolerance limit is reached, the population declines.

Throughout the living world, organisms have become adapted to temperature ranges and fluctuations in a spectacular variety of ways. In the first place, we recognize two broad divisions:

- The "warm-blooded" (*homoiothermic*) forms, the birds and mammals, whose body temperatures are maintained at a uniform level independent of environment; only a few of this group are wholly aquatic,
- The "cold-blooded" (*poikilothermic*) organisms whose body temperatures approximate that of the environment and vary accordingly. Among the latter group are such

extremes as the algae that live in ice water of the polar regions and bacteria that inhabit hot springs at temperatures near 90°C. Between these extremes are found many species of plants and animals variously adapted to thermal conditions.

Great numbers of species are unable to exist as physiologically active stages throughout the range of temperatures experienced in their environment; these typically form spores, or other "resting stages," which exhibit a wider range of tolerance than do the active forms.

Stenothermal and eurythermal species are found in all major taxonomic groups. The European turbellarian, *Crenobia alpina,* is apparently restricted to a temperature regime below 12°C; sessile rotifers are usually found at temperatures above 15°C; tardigrades are active in a wide range from about 0° to 40°C; the optimum temperature for the bryozoan, *Pectinatella,* is between 20° and 23°C, the lower limit of tolerance being near 15°C; the oyster (*Crassostrea virginica*) tolerates temperatures from 4° to 34°C.

We quickly call to mind typical "cold-water" and "warm-water" fishes, many of these being restricted to such environments by the temperature tolerance of the developing eggs; the eggs of the eastern brook trout (*Salvelinus fontinalis*) generally fail to hatch at temperatures above 12°C, the optimum being near 8°C. The optimum temperature for growth of the adult trout is about 15°C. Thus, several optima may be operative in the life history of a given species.

Light appears to be a controlling factor in the distribution of certain animals in time and space. Diurnal vertical migrations of planktonic crustaceans have been noted and studied widely. These movements characteristically take the organisms to near-surface regions during the middle of the night, and to the depths during midday. Numerous explanations have been suggested, and, indeed, the causes may be numerous. However, the correlation between the depth of penetration of the blue segment of the light spectrum and the position of greatest abundance of a copepod species, *Cyclops strenuus,* would indicate a strong influence of the light. Note particularly the conspicuous change in space occupancy with increasing day length from April to June.

From consideration of the "law of the minimum" as it pertains to nutrients, and the "law of tolerance" as it relates to environmental "controls" we can now appreciate the broader *principle of limiting factors*. This principle embodies the concept that population growth and success, in terms of abundance and distribution, are determined by a set of environmental factors, any one of which may, through scarcity *or overabundance*, be limiting.

We have seen a number of examples in which short supply of a requirement could hinder the success of a population; this is not difficult to understand. What is sometimes more difficult to comprehend is that an excess of normally important factors such as nutrient salts or sunlight can exert lethal effects upon organisms. For example, high concentrations of nitrate and phosphate have been shown to be toxic to both plants and animals.

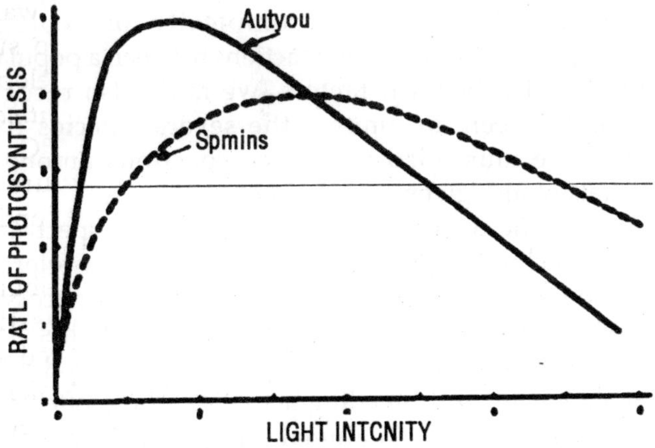

Fig. Photosynthesis (ìmoles CO_2 Absorbed per 10 Iliters of Plant Matter Per hour)

Several exciting aspects of the limiting factors principle can be shown in the reactions between sunlight and phytoplankton in photosynthesis. Note first that in both cases the rate of photosynthesis (measured by rate of uptake of carbon dioxide by plants) is inhibited by "too much" as well as by "too little" light. Light conditions optimums for photosynthesis are indicated at the peak of each curve, but notice that these conditions differ for the two seasonal assemblages. It appears also that the spring phytoplankters exhibit a wider range of tolerance to the light intensities encountered than do those of autumn. Of particular interest is the higher rate of photosynthesis in the dim autumn light than in the relatively bright light of spring. This would seemingly indicate that the autumn community utilizes light more efficiently than do the spring populations. In all probability, however, the species comprising the communities are different.

The principle of limiting factors is not built upon physical and chemical factors alone. Biological features of the environment are often quite influential in determining the abundance and distribution, or over-all success, of a specific population. Once we introduce other organisms and their demands and activities into a set of factors, we immediately perceive greatly increased complexity of ecological relationships. For in addition to reactions between a population and the physicochemical factors, we must also recognize coactions between, or among, the several species living together. We must also recognize coactions among the individuals of any given population.

Interspecific relationships involve a broad spectrum of coactions, including several forms of symbiosis, tolerance, competition for one or many of a great variety of necessities, antibiosis, parasitism, and predation. Within this list of activities are found those that pertain to shelter, food supply and procurement, and breeding activities. These often relate plant to plant, or plant to animal, or animal to animal in a broadly graded series of coactions ranging from those in which one member of the relationship suffers extreme harm at the expense of another, to those in which both of the participants are benefited. Singly, or in combination, interspecific relations may, and often do, act as a limiting factor. They are also basic in the structure and dynamics of communities.

The ability of a species to successfully occupy a place in the economy of a community is also determined by intraspecific relations of the individuals and certain inherent biological qualities of the population. Intraspecific competition for such requirements as food, territory, and breeding partners may serve to limit a given population.

Similarly, overcrowding, which may in some cases lead to stress symptoms, can be limiting in nature. Inherent biological qualities such as frequency of breeding activity, number of offspring produced, survival rate, and mortality rate are often correlated with physiological tolerance to environmental factors. Upon these adaptations depend, at least in part, the abundance and distribution of a population. Some of these population attributes will be examined in the following paragraphs.

Although we have somewhat dissected the set of limiting factors for the sake of illustration, it is important to bear in mind that a single factor can become limiting only when it is near the maximum or minimum of the range of tolerance of a population. In nature, under usual conditions, a given population reacts to a complex of factors. The composition of this complex is not normally fixed; some of the components change seasonally, and a number of interdependencies exist, as we have seen.

A specific organism or population, on the other hand, presents a different story; its tolerance to each factor is more or less firmly established by inheritance and evolution. Thus the position and place of a species in the economy of a community is fairly well defined and delimited by relations between organisms and environmental factors. This thought brings us to another important concept in population ecology, that of the *niche.*

POPULATIONS AND THE ECOLOGICAL NICHE

We have just witnessed how an imposing series of environmental conditions and processes combine to circumscribe the total functions of a given animal or plant population. As a result of evolution of both the complex of environmental factors and the population, the functional reactions and coactions of a particular species have become essentially unique to that species. Thus, each population (characterized by certain rather definite morphological and physiological parameters) and its environmental complex constitute a sort of system, the attributes of which distinguish it from other such systems. Once two systems overlap significantly, the population of each is thrown into competition for nutrient, or other, requirements, and one system breaks down. This "systems" example is a homely analogy to the functional concept of the *niche.*

Can more than one population occupy the same or overlapping niches in natural communities? This question has intrigued ecologists for many years. As yet, the problem has not been solved to the satisfaction of all. The idea is highly important in ecological thinking, however, because, among other things, it offers a point of entry into the study of the factors limiting a single population, and it also relates to events commonly observed in animal communities maintained in the laboratory.

In nature, these forms inhabit small ponds rich in organic matter, and the distributional ranges of the two species overlap. In laboratory cultures, D. *magna* consistently lost in the competition with D. *pulicaria.* Note that in both algae food and the yeast food culture, the population of D. *magna* reached its peak abundance at about 21 days, after which there was a rapid extinction due to interspecific competition. Observe also

the effects under the different nutrient conditions. In algae, the initial growth rate of D. *pulicaria* was quite rapid; in yeast, the rate was less rapid, being exceeded, as shown in the figure, by that of D. *magna.* Two limiting factors, food and oxygen, were operative in this experiment. The results clearly demonstrate that two species do not both persist in a simple environment exhibiting minimal niche diversification. Under natural conditions, of course, food is but one of many factors for which competition by similar species may exist. Intense competition between closely related or ecologically similar species is often avoided by adaptations such as differing feeding times or breeding seasons, or by geographical distribution.

Another important aspect of the niche concept describes the "role" of a species population in the food and energy relationships in a community, i.e., as a "producer" or a "consumer." Although the dynamics and levels involved in these relations will be discussed more thoroughly in the next chapter, it seems worthwhile to mention briefly how these relate to our present topic. The role which a population performs in the economy of a community involves, again, a given set of reactions and coactions.

The role of a diatom, for example, as a producer of the "original" energycontaining substance, is based upon the ability of the plant to carry on photosynthesis. The position, or niche, of the diatom, with respect to solar radiation, raw materials, and herbivorous organisms, is essentially the same throughout time.

Consider, on the other hand, that the role of a great number of animals differs greatly with various stages of life cycles. The important point here is that organisms may be rather narrowly and firmly established in a given niche, or, depending upon adaptations, may function in a number of niches during a life span.

ATTRIBUTES AND DEVELOPMENT OF POPULATIONS

In the foregoing discussions our interest has centered mainly on some of the major physicochemical and biological features of the environment and their effects on populations. How, in the face of such an imposing array of limiting factors, are abundance and integrity of a species population maintained; and, beyond this, how do numbers increase? The answer is found partly in environmental characteristics and partly in the unique attributes of the population.

The essential properties of the population which govern its growth, structure, and spatial distribution are *natality*, *mortality*, and *dispersal*. These qualities of the species are, in a sense, the "counterweights" set against limiting factors, the "degree of balance" militating, as it were, for or against species survival.

NATALITY AND MORTALITY

Natality is a population parameter which describes the rate at which new individuals are produced; it is, in essence, the birth rate. It is often important in certain population problems to consider two aspects of this attribute; these are:

- *Potential natality,* or the maximum rate of population increase if no limiting factors are operative,
- *Realized natality,* or the rate experienced under natural conditions of reactions and coactions.

In estuaries, a population of the American oyster exhibits a very high potential natality, for a single individual is capable of spawning over 100 million eggs. The realized natality, however, amounts to only a small fraction of the potential. Estuarine populations of the sea catfish (*Galeichthys felis*) have a low potential natality; a single female seldom produces more than about 25 or 30 eggs at a time. Parental care, in which the male carries the eggs in his mouth until hatched (oral incubation), contributes to an apparent high realized natality for this species.

The value of the two aspects of natality is found in their usefulness in studying population growth and development in relation to environmental factors and the niche. Potential natality, determined from the rate and number of eggs produced per individual, serves as a standard with which a census of an existing population can be compared. This measure alone, would not, of course, account for the momentary structure of a population. Natality is always a positive quality leading toward continual increase in population size. This effect is essentially offset by population traits such as mortality and dispersal.

Mortality, as a population attribute, pertains to the death rate of individuals of the population. In contrast to natality, mortality has a negative effect on population development, tending to bring about its decline. As in the case of natality, two aspects of mortality may be recognized; these are:

- Physiological longevity, or that experienced under the most favorable set of ecological conditions,
- The realized mortality, or that due to environmental effects prior to fulfillment of the inherited potential.

The realized mortality of a given population can be illustrated clearly by means of a "survivorship" curve derived from plotting the number, or percentage, of individuals still living at successive time intervals against the known life span. If a great majority of individuals of a particular population survived to the limit of the life span of the species, i.e., if physiological longevity were to be experienced, the curve would extend horizontally to physiological longevity and then drop steeply; such a condition is not found in nature.

DISPERSAL

Population dispersal refers to shifting or rearranging of individuals, small groups of individuals, or the entire population with respect to space and to time. Most species possess some adaptation for dispersal at some stage of a life cycle. In some forms, dispersal is accomplished during the seed or spore stage; in others the larval stage may be moved by active swimming or through passive transport by currents or other agents.

Very few natural populations exist in which there is not some immigration and emigration of individuals. The degree of either of these may be quite subtle with respect to numbers and to the rate at which the movement occurs. In this case, dispersal would probably exert little effect on the over-all structure of the aggregation. Conversely, rapid emigration or immigration by large numbers may result in depletion or near-depletion, or in overcrowding of the original stock.

The effects of loss or gain of individuals are important considerations in management of natural populations, such as fishes, for human exploitation. For example, the removal of individuals by fishing of moderate intensity is usually followed by replenishment of the stock. The rate at which the population is rebuilt is, of course, greatly dependent upon the natality-mortality relationship. Extreme exploitation may reduce the numbers below a minimum level necessary for population regrowth. This condition can result in extinction of the population.

Rapid immigration of large numbers of individuals often has deleterious effects on the original population, primarily through increased competition for breeding territory, cover, and food. For many years this idea was not appreciated in fishery management circles, and many bodies of water already at their maximum "carrying capacity" received truckloads of fish fry and/or fingerlings. One result was overpopulation and stunting of all individuals of that particular species; another result was simply that the introduced fishes failed to survive or were eaten by carnivorous inhabitants of the lake or pond.

Under certain conditions natality alone may not be sufficient to maintain observed population numbers. In estuaries, for example, there is a net seaward flour of mixed water sufficient to compensate for the contribution of fresh water by the stream. Thus, estuarine plankton populations are subject to seaward flushing as determined by the rate of water circulation in the estuary.

In order to maintain a given population density, natality, mortality, and dispersal must be attuned to seaward transport. In Great Pond, an Atlantic estuary in Massachusetts, the summer population of copepods (Acartia) has been studied. It was found that the reproduction rates in the uppermost parts of the pond were not adequate to account for the numbers of animals there, that immigration from deeper layers contributed to the population density of the upper region. In the middle porfion of the estuary, natality was apparently sufficient to replace the loss by seaward transport. At the lower end of the estuary, great mortality was experienced, the small population present there being maintained only by immigration of copepods from the upper regions.

As indicated earlier, dispersal, natality, and mortality together constitute a set of interacting factors which essentially regulate growth and structural dimensions of a population. We have seen that there is a tendency for populations to grow, due to the fact that the potential natality normally exceeds the potential mortality. However, the extent to which such growth proceeds would be tempered by dispersal traits. Should the realized natality exceed the realized mortality under conditions of minimum dispersal activities, the population would increase. Under the same natality-mortality conditions, and with excessive emigration, the population could be expected to decrease.

With equal birth and death rates, the population would tend toward stability, but again, this is influenced by movement of individuals. If the realized mortality exceeds the realized natality, extinction of that population would be expected; recruitment from nearby aggregations of the same species could serve to maintain the "unbalanced" population. Having given attention to these "forces" which mold population growth and density, let us now turn to consideration of these two population parameters.

GROWTH OF POPULATIONS

All natural species populations possess an *innate capacity for increase* at a maximum rate—the *biotic potential.* For a given species, this maximum rate is set by inherited potential natality-potential mortality values. These values differ greatly with species. Compare, for example, the figure given previously for a single oyster, with that of man. In nature, the biotic potential is not normally fulfilled, due to the limiting features of many environmental factors.

Thus, opposing the innate tendency of a population to increase is a set of conditions which, in effect, offer resistance to population growth. We recognize, therefore, that the growth and success of a species population depend upon the degree of "harmony" struck between biotic potential and *environmental resistance.*

The actual, or realized, rate of growth of a population is determined by the initial number of individuals, plus additions, minus the number lost per unit time. If, during a period of one year, two individuals produce 24 offspring from which four are lost, the population density would stand at 22 at the beginning of the following year. If similar conditions of natality and mortality prevail during the second year, the census should be 242 individuals. In other words, the rate of population growth is geometric, and is represented by the relationship:

$$\frac{dN}{dt} = rN$$

in which N is number of individuals and t is time.

The symbol r represents the rate of natural increase of a population, in other words, the difference between natality and mortality at a given moment;

$$\frac{dN}{dt}$$

gives the average rate of change of N per unit time. If such growth continues unlimited, it takes the form indicated by the curve in Figure.

Since a particular community can support only a given number in any one species population, it is apparent that the maximum biotic potential can not be maintained in nature. Consequently, the form of the growth curve of populations subject to limiting factors must differ from the "unlimited" curve in Figure, and so it does. As a population begins to increase under favorable environmental conditions, growth is typically rapid and passes through a *phase of accelerating growth*.

This is especially characteristic of populations exhibiting high natality and under conditions of little intraspecific competition. In time, the rapidly increasing numbers do create problems of maintenance, and the growth rate begins to slacken, the population passing through a phase of *decelerating growth*. The point on the curve at which the rate of increase slackens is called the *inflection point*. Eventually the population size becomes somewhat stabilized through essential equilibrium of reactions and coactions.

This size level may be at, or near, the *carrying capacity* of the community. On the curve, the carrying capacity (number of individuals) is designated the *asymptote*. Note that the curve representing population growth under natural conditions is typically in the form of a shallow "S," or the *logistic curve*.

Given selfregulating growth processes, the curve may be derived from the differential equation:

$$\frac{dN}{dt} = rN\frac{(K-N)}{K} \text{ or } \frac{dN}{dT} = rN\left(1-\frac{N}{K}\right)$$

where τ is the biotic potential for each individual in the population; N is the total momentary population size; t is time; and K is the maximum number of individuals possible in the population under prevailing conditions. The expression,

$$\frac{(K-N)}{K}$$

describes the extent of resource utilization by increasing population; in other words, the growth of the population itself causes the environment to become more limiting. As a result of this condition, the potential rate of reproduction is decreased as the population size grows toward the asymptote. This equation tells us that the population growth rate is equal to the maximum possible rate of increase times the degree of realization of that maximum potential rate.

The equilibrium which a population maintains about the asymptote is primarily a function of natality (*n*) and mortality (*m*), and, under certain conditions, dispersal. At equilibrium, n - m = 0, meaning that losses due to mortality are offset by additions derived from births. Deviations above the asymptote occur when *n* - *m* is positive (population increase); deviations below the asymptote follow when *n* - *m* is negative (population decrease).

In the growth of a laboratory population of a small fish, the guppy (*Lebistes reticulatus*), are fitted to the logistic curve. Note the characteristic phases of growth until the asymptote is attained. Observe also that the population increase initially "overshoots" the asymptote, but returns to the carrying capacity; this phenomenon is often encountered in population studies.

Upon reaching its asymptote, a population does not normally remain at uniform density, but rather fluctuates in numbers in response to one or more ecological or inherited factors. Irregular fluctuations, or relatively asymmetrical departures from equilibrium, may result from variations in any critically limiting environmental factor. A great number of observations on populations of plants and animals have shown symmetrical fluctuations, or oscillations, of population numbers in response to factors regulated by daily rhythms, tidal cycles, seasonal and annual cycles, and to predator-prey and host-parasite relationships.

Within a given population, the relative number of individuals in each age level is important in determining the growth form, as well as serving as a basis of diagnosis of the condition of the population. These general diagnoses assume that the distribution of age groups in the population is stable. The importance of age distribution in a population lies in the cumulative effect of the natality-mortality values for each of the age classes on the general population. As indicated above, stability of population structure depends upon nearly equal rates of made reference to an oligochaete population with a density of 19,000 individuals per square meter of bottom surface in the saline region of an estuary.

Population density is also time-relative in that the number, or mass, of individuals may, and often does, vary temporally. Daily vertical movements of certain zooplankton, for example, result in changes in population density at a given level in a lake; seasonal migrations of some fishes and other animal cause fluctuations in population size. Certainly, marked changes in population density are observed during breeding times.

An observed population size represents the product of a number of environmental features and relationships working through biological processes. Thus, analysis and study of density yield mush enlightening information on growth form and the effects of specific environmental factors on population development. It is often difficult, however, to recognize and measure subtle effects operative under natural conditions. Ecologists frequently surmount such difficulties by laboratory studies under controlled conditions, and by the judicious use of statistics to test the meaning and validity of laboratory and field observations.

Although influenced to a considerable extent by biological attributes of the species and the biogeochemical features of the environment, population density, per se, exerts effects on the members of the population and upon the environment. The density of a given population in a prescribed space may affect the population rate of increase, the proportion of males and females, the growth rate of individuals, the rate of utilization of nutrient substances and respiratory gases, and other processes.

Thus, there is an optimal density level for the growth or success of a population, and this implies that there must also be proportional limitation of growth or success under overcrowded or undercrowded conditions. It has long been known that there are deleterious effects on growth, reproduction, survival, and other biological processes which accompany overcrowded situations.

It remained for W. C. Allee and his students to demonstrate that at all levels of the animal kingdom there is added safety in numbers up to the optimal population level; there are adverse effects accruing to undercrowded populations. This is known as the *Allee Effect*. One example of this obtained by noting, the relative rate of population increase. If data from a logistic population curve are replotted to show the change in rate of increase (*dN/dt*) against density, one can see that at the low and high densities there is a low rate of increase while the highest rate of increase occurs at some optimal population level.

There may be several mechanisms whereby aggregations of individuals can grow and survive better than isolated individuals in the same medium. One such mechanism involves decrease of exposed surface. In the group, there is less surface area per individual exposed to the environment than would be if the individuals were dispersed. For example, the individuals in a group of flatworms have less total surface exposed to adverse effects of ultraviolet light or other extreme environmental factors than would be the case if these individuals were dispersed and exposed.

Similarly, masses of sea-urchin sperm survive loner than dispersed cells since there is less environmental exposure to the sperm inside the mass. It might be noted, also, that there are undercrowding effects on the embryonic development of sea-urchin eggs.

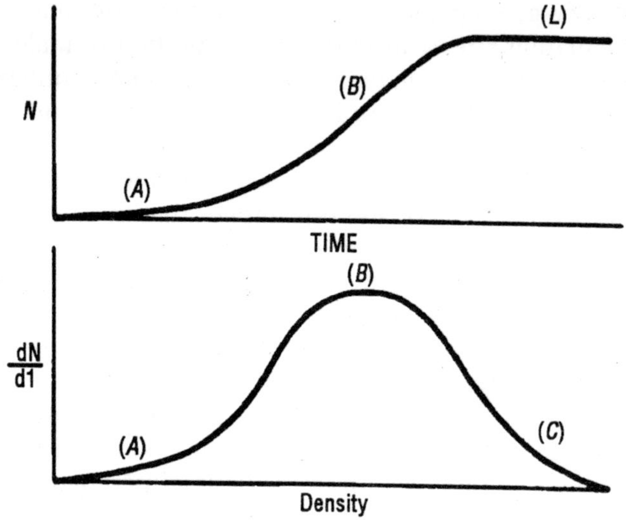

Fig. Schematic Representation of the Allee Effect, in which data from the logistic curve are plotted to depict rate of population change (*dN/dt*) as a function of density. According to the lower figure, both low density and high density may inhibit rate of population growth.

Another aspect in the success of groups versus individuals is that the group can more efficiently condition the medium by removing poisons, changing the salinity of the water, or by adding growth-promoting factors. It has been demonstrated that groups of goldfish can survive longer than isolated individuals when placed in media poisoned by silver nitrate; the group more quickly removes and detoxifies the poisons, whereas individuals usually die.

The groups of goldfish also grow faster as a result of sharing particulate food that spews from their mouths, and as a result of a higher concentration of a growth factor associated with the slime produced by the fish. Other studies have demonstrated that groups of protozoa have a higher per-individual growth rate than do isolated individuals due to the greater production of some growth-promoting factor, and also because the group can better control the population of bacteria.

There may also be benefits to the group resulting from more efficient coordination of activities that produce greater synchrony in reproduction. This heightened psychological state that leads to greater sychronization of activities and consequently to less mortality, has been shown for shore-bird colonies.

DISTRIBUTION OF INDIVIDUALS OF A POPULATION

We have previously considered dispersal of population individuals as a process and condition contributing to population growth form. Of utmost importance to population dynamics generally, is the manner, or pattern, in which members of a population are distributed within the community. This quality of populations is also of primary concern to the study of intrapopulation relationships.

Within the space occupied by a given population the individuals may become dispersed in a *random* pattern, in an *evenly-spaced* pattern, or as *aggregations*. Obviously, the pattern exhibited by a given population is subject to considerable variation depending upon the census of the population, the number of sites available for occupancy, food accessibility, and a host of other factors. Equally important in setting the dispersal pattern is the inherent sociological nature of the species at various seasons. Important as the problem of animal distribution is, it has received insufficient attention, especially as pertains to aquatic organisms. Future researches may well reveal additional patterns of intrapopulation distribution.

Completely random distribution of individuals is apparently uncommon in nature. The degree of "randomness" in the dispersal of individuals is determined statistically by several methods. One of these involves the fitting of observed distributional data to the "Poisson series," a statistical measure of the frequency with which aggregations of 0, 1, 2, 3, 4, 5... n can be expected in plots or quadrats. Deviation from the Poisson indicates various patterns of nonrandom distribution.

It has been suggested that the lack of random dispersals in nature results from a generally inherited tendency of most animals to aggregate to some degree.

Evenly spaced distribution patterns are probably best explained on the basis of tendencies of many species toward *territoriality*. As an animal trait, territoriality refers to defence of a given area by overt aggressive action. The rather euryhaline cyprinodont fish, *Cyprinodon variegatus,* exhibits marked territoriality during breeding season. At this time, the male fish come to occupy vigorously defended areas in the shallow zone of marsh streams. The territories are small, usually being about 30 to 50 cm in diameter, and the aggressive behaviour lasts for several days; any male intruder is immediately attacked by a territory holder.

The females are usually somewhat concentrated a short distance from the area of territories, and show little excitement. With considerable frequency females singly enter the territory of a male, where a brief breeding encounter takes place. Territoriality is often associated with dominance-subordinance levels developed in many animals. Green sunfish (*Lepomis cyanellus*) in a laboratory colony develop hierarchies and defend their respective positions. Depending upon numbers of individuals and available territories, the most subordinate fish may not be able to establish a place in a "peckorder" such as do members of chicken flocks.

The *home range* of an animal is the region in which the individual normally travels. The important distinction between territoriality and home range lies in the aggressive response to invasion of the area shown in territoriality; home range carries no connotation of aggressive behaviour. The concept of home range has been investigated less in aquatic animals than in terrestrial forms. However, a fair volume of literature on the movement of tagged fishes in lakes and streams strongly points to homing and home range in these animals.

For example, of 4557 fishes marked and released at Grapevine Point, Douglas Lake, Michigan, from 1937 to 1939, not a single one was caught from any other point.

Investigations in an Indiana stream revealed little movement of fishes between pools. Most recently, results from a study of homing in the newt (*Taricha* (= *Triturus*) *rivularis*) in a California stream convincingly attest to observance of home range in this semiaquatic species. Even blinded individuals returned to their original locality after having been displaced some 0.8 km downstream.

The tendency to form aggregations is widespread in both plants and animals, and results in the most frequently encountered pattern of distribution. These aggregations, however, may be somewhat randomly scattered over a given space. Aggregation has been described as "a uniform response to a nonuniform environment"; it may also be a response to inherited social traits. One can call to mind a wide array of aggregation patterns, ranging from pronounced "schools" of certain fishes to tightly formed clumps of flatworms on the undersurface of a stone on a stream floor.

Time and space do not permit us to give due consideration to the many aggregation patterns. As to causes of this general type of distribution, we can list: annual, seasonal, or nocturnal-diurnal variations in physicochemical qualities of the environment; coactions among biological components of the community; physiological processes such as reproduction; and the above-mentioned sociological attributes of the species. The effects of aggregative tendencies are varied and manifold; some of these have been presented in the preceding discussion of population density.

THE SPECIES, UNIQUELY UNIQUE

Throughout history and in a variety of contexts, the question has come up: Just what is the human species? A notoriously superficial answer to what a human is relates to the elements of which the human body is composed. Like all organisms, we are composed mostly of carbon, nitrogen, hydrogen, and oxygen, with the last element constituting nearly two thirds of our total weight. In fact, if sodium, calcium, phosphorus, potassium, chlorine, sulpher, and magnesium are added, 99.9% of our constituents have been taken into account. An array of heavy metals from aluminum to zinc represents the remaining minute percentage. This litany of chemicals, however, is not much of a start on any definition of this species.

By definition, every species is unique, but is there something about *Homo sapienssapiens* that actually makes this species *uniquely* unique among species? Many of the contexts in which this question arises are well outside the scope of a quasi-scientific ecology text. But two contexts are particularly relevant: the scientific and the political. Ecologically, the human species occurs as one or many populations distributed throughout the globe and occupying the most flexible niche of any species. Politically, the human species is perceived as a particularly devastating organism potentially capable of destroying whole ecosystems and perhaps the biosphere itself.

We may be unique in our ability to adapt to an extraordinary range of environmental conditions, not through physiological or other "purely" biological means but through intelligence and our ability to communicate abstract ideas, including concepts of the past and future. We can learn from the past of our species as well as from personal experience. We can plan for the future, often by applying past history. Language, writing, culture, and both an individual and a political sense of morality may be features that render our species quantitatively and qualitatively unique in a sense no other species is unique. Our minds represent a new level of organization as profound a leap from life as life is from physicochemistry.

Human horizons are constantly expanding, literally scaling new heights and exploring new dimensions. We see ourselves as Lilliputians, poised precariously on a very small planet. But through intellect, we "have the extraordinary ability to examine and comprehend the structure of the universe. This [alone] truly distinguishes us from less-evolved animals." The potency of the human mind is one lesson from contemporary astronomy.

With this recognition combined with concepts of the Gaia Hypothesis or of deep ecology comes a paradox for all of us but most particularly for environmental professionals. If our species is a part of nature, our actions cannot be other than natural, no matter how detrimental to our ecosystems. If our cultural organization distinguishes us from other life forms, how can we be bound exclusively to follow "nature's way"?

In the end, it would seem that evolution has led to human nature, which has led to culture; culture has enabled the behaviour that is believed by many to impair nature's integrity. We face a quandary in that any return to nature for us would contravene our nature and our history.

In seeking a way out of our quandary, it is worth contemplating the ecologically sound assertion placing populations beyond good and evil. It is the nature of every population to grow to fill its ecosystem. When too numerous, organisms either die out or transcend themselves. Those that transcend "find new ways to procure room, carbon, energy, and water." But they produce new wastes, which test them. At what stage is the human population? *Are* we too numerous? If so, are we at the brink of extinction? Or are we about to transcend ourselves?

THE POPULATION ECOLOGY OF THE HUMAN SPECIES

At least one ecologist holds the human species unique in our capacity to alter our niche seemingly at will. We can increase the sources of niches in deriving energy from something other than complex organics we ingest, and we can use novel materials and systems unlike any other species. Another scientist finds the combination of size of brain, social structure, speech, and the use of tools to be what sets humans apart.

In other words, the biological system we call culture is unique to us and a new level of organization. It is culture that allows us to manipulate our environment in ways that are both purposeful and anticipatory.

It is in culture that our unique capacity for abstract reasoning—from mathematics to aesthetics-comes to its culmination. The strategy of human cultures, each a suprapopulation, is to improve the quality of life for one or more future generations.

Sometimes our biological nature and our cultural one are in conflict. Biologically, it was expedient when some thousands of years ago we attained the capacity for which every species is striving: the ability to alter the environment to eliminate other species. Morally, we are beginning to take a second look. Doing so does not come easily. Has Garrett Hardin's "tragedy of the commons" placed us in need of "lifeboat ethics"? Must some individuals be denied a place on the global lifeboat for the sake of those already on board?

Indeed, we may not yet have escaped our first niche. We started out as wandering ice age foragers, and our physiques, temperaments, desires, and patterns still reveal that beginning. There are those who consider humans either an unnatural population or, perhaps more accurately, a supranatural population. There are others who perceive urban human aggregations as akin to a climax community. Whatever the scientific reality and whatever the political, economic, and social ramifications of those realities, *Homo sapiens sapiens* is a particularly important biological system to environmental professionals.

If one is to analyse humanity in ecological terms, one of the first questions to arise is whether the species is subject to limiting factors. Or does that seemingly infinite capacity to adapt into expanding ranges of tolerance, through seemingly infinite expansions of consumption, have a maximum? Maximum or not, can human populations thrive under near-optimum conditions without compromising either the carrying capacity of their respective ecosystems and ultimately the biosphere itself? Are we really jeopardizing other species in some "unnatural" sense, or are we merely manifesting the traits of all species in filling and expanding our niche?

Significantly enough, we may be the only species capable of asking such questions and moved by a sense of obligation associated with the answers.

As Odum has observed, we must *transcend* the principles of general ecology because of our unique abilities to control our immediate surroundings, and our tendency to develop culture independently of environment.

First, we must recognize that human populations are parts of widely ranging biotic communities and ecosystems. Superimposed on that status is the reality that our ecology goes beyond mere demographics—the statistical functions of any population. Causes and effects driving our populations demand closer analysis than statistics alone can provide. Finally, internal dynamics as well as relationships to external factors must be taken into account. Where biological regulation may not succeed, cultural regulation may be demanded.

We are responsible for any number of perturbations in ecosystems. Among the most frequently recognized are nutrient enrichment, acidification, toxification, nutrient depletion, habitat modification and outright destruction, and both the elimination of native species and the introduction of alien species. Above and beyond assumption of our niches, our intrusions into ecosystems can be stimulatory or inhibitory or can impose structural alteration.

In considering the ecology of human populations for purposes of application, most particularly in the marketplaces and in law and policy, it is essential to expand the questions and any answers derived beyond ecology into the cultural fields that do make us "uniquely unique." When we perceive anthropogenic ecological perturbation, the first step is to comprehend the ecological results of that perturbation. But before we can turn to cultural institutions to prevent consequent harms or to protect or restore the ecosystem, we need to explore the cultural implications of the perturbation, of its ecological results, and of available responses. Only then can we begin to take effective action.

When we do act, we must always remain sensitive to ecological complexity. We must heed the lessons of past intrusions into ecosystems, whether taking the form of intervention or representing a perturbation. The former is presumably, but not necessarily, beneficial to the system, the latter detrimental. The truth is, we seldom are able to predict whether an action will be intervention, intrusion, or perturbation. Too often, whatever the intent or lack of it, our actions result in ecological backlashes. These are unexpected and certainly unintended ecological responses. In simple terms, ecological backlashes can be categorized as failures to secure the intended correction or as additional perturbation in the subject ecosystem or one somehow connected to it.

In short, environmental professionals and, indeed, the population at large must heed all the ecological principles and concepts described here. We must take them personally! Only then are we in a position to apply our cultural selves to resolving our place in the biosphere.

SYSTEMS ECOLOGY AND RESOURCES

In seeking to apply ecological principles to human activities, some rhetorical questions should lie behind those efforts. What are the practical implications of the phenomenon that the optimum for quality is always less than the maximum population that an ecosystem can sustain? To what extent are deteriorating environments attributable to cultural tendencies to be too independent of the natural environment?

Systems ecology, emergent with the analytical power provided by computer technology, is a kind of return to the black box described in the opening chapters. When attempting to analyse the ecosystem or some component thereof, the ecologist or the environmental professional is immediately confounded by the sheer complexity of the interactions at and among all the relevant levels of organization. As should be thoroughly instilled by now, every level of organization from pure mathematics to planetary ecology and beyond has something to contribute to an ecosystem's output as a result of each and every isolated input. In a very real sense, systems ecology and computer modeling are founded in the Theory of Integrative Levels.

Systems analysis is defined for purposes of ecological systems as the process of translating internal physical or biological concepts into a set of mathematical relationships that are then manipulated to predict or explain the behaviour of that system. Some of the complexity can be transformed into the mathematical symbols and equations on which computer modeling depends brief, the mathematical symbols can be a useful shorthand describing complex ecological systems. The equations can permit a formal statement (which can often be transformed into a highly effective visual presentation) of how an ecosystem's components are likely to react to change.

The model resulting from the mathematics derived through systems analysis remains a mathematical system. However valuable, it will always present an imperfect and abstract simplification of the real world. The challenge, of course, is to work with that relatively simplistic representation while recognizing its imperfections in order to serve the ecosystems upon which we depend for life. As every research scientist can avow, even tentative answers open the way to a fuller understanding of how the natural world functions, allowing for predictions and some degree of control.

But it is the real world that is correct, if there is any conflict between the answers and the underlying reality reflected in the information provided by the natural sciences. Wrong answers and other "failures" serve to reveal flaws in the conceptual framework and may point the way to corrections in the modeling process. If applied to a living system, however, wrong answers can do more harm than good, intentions notwithstanding.

The criteria to be applied in evaluating models are threefold: realism, precision, and generality. Realism is the degree of correlation with the biological concepts represented by the model. Precision is expressed through the ability to predict quantitative change and to mimic the data upon which the model is based. Generality is the breadth of applications the model fits. Models are also characterized in terms of resolution and wholeness. The former expresses the number of attributes included in the modeling process; the latter, the interactions recognized between levels of organization.

Allegedly, ecological models are noted more for their generality than for their precision. There is an argument to the effect that a model cannot at once be general and precise. Any improvement in the one is matched by a loss in the other.

Moreover, an ecological model should be expected to be unrealistic in terms of lower levels of organization because of the lack of complete information concerning input and output at those levels, the sum of which are ultimately being expressed at the level of interest.

Whatever the tools of the environmental professionals, it is important to remind ourselves constantly to expand our thinking and to take care in imposing restrictions in that technology alone is not enough. "Moral, economic, and legal constraints arising from full and complete public awareness that man and the landscape are a whole must also become effective." Environmental professional or private citizen or both, it is essential further to heed the ecological principle that the optimum for quality for a population is always less than the maximum quantity that can be sustained.

Another ecological principle that must be heeded is that the more we demand of an ecosystem, large or small, the less energy is retained by that system for maintenance. Has *Homo sapiens sapiens* been justifiably labeled a parasite on the environment at large? Are we truly as destructive to our host as most parasites?

HUMAN NATURE ENVIRONMENTAL PHILOSOPHIES

Throughout the history of the human species, environmental philosophy has played a major role. That role has, however, undergone profound alteration, especially in Western civilization as it is manifesting itself and merging with others here in the United States.

We have gone from huntergatherer to tiller of the soil to industrialist to scientist to the age of information. Both our niche and our intellectual relationship to our ecosystems have altered accordingly. Emerging from the status of one more primate species in the wild, we have learned to fear the natural world. It is in our nature to seek conquest of whatever we fear.

Some may believe human populations have conquered the natural world, but most of us realise that our place as conqueror is tenuous at best. In recent years, we have come to realise that we may represent a unique hazard not only to our ecosystems but to the biosphere as a whole. We are rediscovering our niche and the habitats in which it is active.

When humans left the forests and fields to enter into agricultural and later societies, they came to fear the wilderness. When North America was invaded by representatives of Western civilization, the frank intent was to conquer those usurping the land—wild beasts, howling wilderness, and hellish fiends. This vast wilderness was a place of evil as well as one of physical danger. Conquest was a moral imperative.

The contemporary environmental movement in the United States traces its most recent roots to the conservation and preservation movements in the early twentieth century. In fact, early conservation helped shape the notion of "commons" as a public domain in resources that are held not susceptible to private appropriation. Conservation seeks to ensure the preservation of a quality environment in company with a continuous yield of useful plant, animal, and natural materials—natural resources. Conservation, then, tends to be more anthropocentric than preservation, which tends more toward biocentrism or ecocentrism.

Our literature and the popular arts in the United States are rich in images identifiable as personal ideas of wilderness, conservation, and preservation. Among the best known of the writers is Aldo Leopold, who spoke of conservation as a "state of harmony between men and land"—his "land ethic," an ecological conscience. Ecological health, according to Leopold, lies in the land's capacity for self-renewal. Conservation arises from efforts to understand that capacity in order to preserve it. The land ethic is a major concept incorporated into the contemporary environmental movement.

THE CONTEMPORARY ENVIRONMENTAL MOVEMENT

What Is It? In the early 1960s, the environmental movement in the United States still took the form of conservation or preservation. In the 1970s, the shock of gross pollution and disappearing landscapes moved us to a kind of grim pleasure expressed in part as cultural and individual ambivalence toward progress. The environmental movement provided one outlet for our malaise. Whatever our individual relationship to the movement, we can all probably agree that the role of culture is not to overcome nature but to be open to its potential. A primary lesson from ecology is that all "poverty" is attributable to unrestrained population growth. The environmental movement, then, is seeking ways to alleviate the results of population growth, unrestrained at least at the local level, and thus to relieve ecological poverty in whatever biological systems in which it is being expressed.

An Ecological Perspective

To the extent Odum can be cited for the ecological perspective, he has observed that ecologically sound conservation, one of the missions of contemporary environmental policy, is neither a matter of hoarding resources nor one of total rejection of development. Instead, the objective should be one of rejection of unplanned development in violation of ecological laws. In the past, such ecologically unsound development has rightfully been attributed to ignorance as well as to shortsightedness and advancement of parochial interests.

Today, there is no excuse for development characterized by any of those three factors. In the United States, at least, there is a very real movement toward integration of all the factors supporting development that is ecologically responsible. It is a youthful movement, however, and subject to all the excesses of youth. At least we are moving away from all three impediments to environmental quality.

One of the major problems, and one that may be insurmountable, is the nearly total focus on the human perspective. We tend to speak in terms of environmental quality in association with products, recreation, and aesthetics. We seek to ensure continuous yields of useful plants, animals, and materials. Not until we act on the knowledge that ultimately it is only healthy environments that are healthful are we likely to change our ways in an ecologically sound manner.

As environmental professionals seek to serve what is sometimes called stewardship of ecosystems and of the biosphere itself, some turn to cultures not ordinarily included in that of mainstream America. For example, some assert that we should heed the injunction of certain Native Americans to consider the effects of our actions through seven generations. All too often, we have barely thought beyond the current generation. Another lesson to take from Native American peoples is the notion that we cannot preserve ecological systems by the sole expedient of dictating by law. There instead must be a profound alteration in the way each of us relates to the world around us.

Chapter 6

Waterborne Diseases

Water-borne diseases are any illness caused by drinking water contaminated by human or animal faeces, which contain pathogenic microorganisms. The full picture of water-associated diseases is complex for a number of reasons. Over the past decades, the picture of water-related human health issues has become increasingly comprehensive, with the emergence of new water-related infection diseases and the re-emergence of ones already known.

Data are available for some water-, sanitation- and hygiene-related diseases (which include salmonellosis, cholera, shigellosis), but for others such malaria, schistosomiasis or the most modern infections such legionellosis or SARS CoV the analyses remain to be done. The burden of several disease groups can only partly be attributed to water determinants. Even where water plays an essential role in the ecology of diseases, it may be hard to pinpoint the relative importance of aquatic components of the local ecosystems.

ANAEMIA

Anaemia is a lack of red blood cells and/or haemoglobin.

This results in a reduced ability of blood to transfer oxygen to the tissues. In the milder form anaemia is asymptomatic. In the more severe form it is associated with fatigue, weakness, dizziness and drowsiness. Signs include loss of normal colour in the skin and in the lips, tongue, nail beds and the blood vessels in the white of the eyes.

Without treatment, anaemia can worsen and become an underlying cause of chronic health, such as impaired foetal development during pregnancy, delayed cognitive development and increased risk of infection in young children.

Anaemia is common throughout the world. Its main cause, iron deficiency, is the most prevalent nutritional deficiency in the world. Several infections related to hygiene, sanitation, safe water and water management are significant contributors to anaemia in addition to iron deficiency. Nine out of ten anaemia sufferers live in developing countries, about 2 billion people suffer from anaemia and an even larger number of people present iron deficiency.

ARSENICOSIS

Arsenicosis is a chronic illness resulting from drinking water with high levels of arsenic over a long period of time (such as from 5 to 20 years). It is also known as arsenic poisoning. The WHO recommends a limit of 0.01 mg/l of arsenic in drinking water. It results in various health effects including skin problems, skin cancer, cancers of the bladder, kidney and lung, and diseases of the blood vessels of the legs and feet, and possibly also diabetes, high blood pressure and reproductive disorders.

Natural arsenic contamination is a cause for concern in many countries of the world including Argentina, Bangladesh, Chile, China, India, Mexico, Thailand, and US. Because of the delayed health effects, poor reporting, and low levels of awareness in some communities, the extent of the adverse health problems caused by arsenic in drinking water is unclear and not well documented.

A SCARIASIS

Ascariasis is an infection of the small intestine caused by the roundworm *Ascaris lumbricoides*. As larval stages travel through the body, they may cause visceral damage, peritonitis and inflammation, enlargement of the liver or spleen, toxicity, and pneumonia.

A heavy worm infestation may cause nutritional deficiency; other complications, sometimes fatal, include obstruction of the bowel by a bolus of worms (observed particularly in children), obstruction of bile or pancreatic duct. Children are infected more often than adults, the most common age group being 3-8 years. The infection is likely to be more serious if the nutrition is poor. Eating uncooked food grown in contaminated soil or irrigated with inadequately treated wastewater is a frequent avenue of infection.

Roughly 1.5 billion individuals are infected with this worm. The disease is found worldwide, with greatest frequency in tropical and subtropical regions, and in any areas with inadequate sanitation.

BOTULISM

Botulism is an acute poisoning resulting from ingestion of food containing toxins produced by the bacillus *Clostridium botulinum*. This bacterium can grow only in an anaerobic atmosphere, such as that found in canned foods; botulism is then almost always caused by preserved foods that have been improperly processed. Person to person transmission of botulism does not occur.

Classic symptoms of botulism occur between 12-36 hours after uptake of the botulinum toxin, but they can occur as early as 6 hours or as late as 10 days. They are not caused by the organism itself, but by the toxin that the bacterium releases. Incidence of botulism is low, but the mortality rate is high if treatment is not immediate and proper. The disease can be fatal in 5 to 10% of cases.

The symptoms usually include dry mouth, difficulty swallowing, slurred speech, muscle weakness, double vision, vertigo, vomiting, and severe diarrhoea, along with a progressive muscle paralysis. If untreated, these symptoms may progress to cause paralysis of the arms, legs, trunk, and respiratory muscles. There is no fever and no loss of consciousness.

Botulism is a relatively rare disease. Worldwide, food-borne botulism is the most commonly reported form, but the United States experiences a higher incidence of infant botulism. In the US an average of 110 cases of botulism are reported each year. Infant botulism is possibly under-reported worldwide because it is a relatively recent discovery and diagnosis can be difficult.

CAMPYLOBACTERIOSIS

Disease caused by *Campylobacter jejuni*, now the leading cause of bacterial food poisoning, most often spread by contact with raw or undercooked poultry. A common way to become infected is to cut poultry meat on a cutting board, and then use the unwashed cutting board or utensil to prepare vegetables or other raw or lightly cooked foods. The Campylobacter organisms from the raw meat can then spread to the other foods. The organism is not usually spread from person to person, but this can happen if the infected person is a small child.

Symptoms of the infection include diarrhoea, abdominal pain, malaise, fever, nausea and vomiting. The illness usually lasts 2 to 5 days but may be prolonged by relapses. Rare complications include seizures due to high fever or neurogical disorders. Death from Campylobacteriosis is rare and it is more likely in the very young, the very old, or those already suffering from a serious disease such as AIDS.

This disease occurs worldwide. Sanitation, personal and food hygiene as well as safe water supply are important in its prevention.

CHOLERA

Cholera is an acute, diarrhoeal illness caused by infection of the intestine with the bacterium *Vibrio cholerae*. A person may get cholera by drinking water or eating food contaminated with the cholera bacterium. The disease can spread rapidly in areas with inadequate treatment of sewage and drinking water. The cholera bacterium may also live in the environment in brackish rivers and coastal waters.

Approximately 1 in 20 infected people has severe disease characterized by profuse watery diarrhoea, vomiting and leg cramps. In these people rapid loss of body fluids leads to dehydration and shoal. Without treatment death can occur within hours.

Cholera cases and deaths were officially reported by WHO, in the year 2000, from 27 countries in Africa, 9 countries in Latin America, 13 countries in Asia, 2 countries in Europe, and 4 countries in Oceania. In the same year some 140,000 cases resulting in approximately 5000 deaths were officially notified at WHO. Africa accounted for 87% of these cases.

CRYPTOSPORIODIOSIS

Cryptosporiodiosis is a disease affecting the intestines of mammals that is caused by Cryptosporidium, a protozoan parasite in the *phylum Apicomplexa*. It is spread through the faecal-oral route.

Symptoms appear from two to ten days after infection and last for up to two weeks or so. As well as watery diarrhoea there is often stomach pains or cramps and a low fever. Some individuals are asymptomatic but are still infective and thus can pass on the infection to others. Even after symptoms have finally subsided that individual is still infective for some weeks.

Despite not being identified until 1976 it is one of the most common waterborne diseases and is found worldwide.

CYANOBACTERIAL TOXINS

Cyanobacterium is a photosynthetic bacterium of the class Coccogoneae or Hormogoneae, generally blue-green in colour and in some species capable of nitrogen fixation. It is also called blue-green alga. Some species of cyanobacteria produce toxins that effect animals and humans. The disease due to it varies according to the type of toxin and the type of water or water-related exposure. Humans are affected with a range of symptoms including skin irritation, stomach cramps, vomiting, nausea, diarrhoea, fever, sore throat, headache, muscle and joint pain, blisters of the mouth and liver damage.

People are mainly exposed to cyanobacterial toxins by drinking or bathing in contaminated water. Some species form a scum on the water. Surface scum, when they occur, represent a specific hazard to human health because of their particularly high toxin contact.

Cyanobacterial toxins in lakes, ponds, and dugouts in various part of the world have been known to cause poisoning in animals and humans.

They have been linked to illness in various regions throughout the world, including North and South America, Africa, Australia, Europe, Scandinavia and China. There are no reliable figures for the number of people affected worldwide.

DENGUE

Dengue is an acute infectious disease caused by a virus and transmitted by the bite of the Aedes mosquito, also known as break bone fever and bone-crusher disease. Dengue fever is a severe, flu-like illness that affects infants, young children and adults but rarely causes death.

The symptoms, following an incubation period of five to eight days, are fever, chills, and severe headache, pain in the joints, sweating, and prostration. The clinical features vary according to the age of the patient.

Infants and young children may have a feverish illness with rash. Older children and adults may have either a mild feverish illness, or the classical incapacitating disease with abrupt onset and high fever, severe headache, pain behind the eyes, muscle and joint pains and rash.

Dengue hemorrhagic is a potentially lethal complication and is today a leading cause of childhood death in several Asian countries. It is characterized by high fever, haemorrhage, often enlargement of the liver, and in the most severe cases circulatory failure.

Today, dengue is found in tropical and subtropical regions around the world, predominately in urban and per urban areas, where Aedes mosquitoes are prevalent. Globally there are an estimated 50-100 million cases of dengue fever each year.

DIARRHOEA

Diarrhoea is frequent discharge of watery faeces from the intestines, sometimes containing blood and mucus.

Diarrhoea due to infection may last a few days, or several weeks, as in persistent diarrhoea. Persistent diarrhoea may result in severe dehydration and shock. It is therefore necessary to replace the fluid lost by the body. Severe diarrhoea may be life threatening due to fluid loss in watery diarrhoea, particularly in infants and young children, the malnourished and people with impaired immunity.

Diarrhoea is a concomitant of many infectious diseases, especially typhoid fever, bacillary or amoebic dysentery, and cholera. Diarrhoea is a symptom of infection by a host of bacterial, viral and parasitic organisms most of which can be spread by contaminated water.

It is a rare occurrence for most people who live in developed countries where sanitation is widely available, access to safe water is wide and personal and domestic hygiene is relatively good. Diarrhoea due to infection is widespread throughout the developing world. In Southeast Asia and Africa it is responsible for 8.5% and 7.7% of all deaths respectively. In 1998 it was estimated to have killed 2.2 million people, most of whom where under 5 years old.

DRACUNCULIASIS OR GUINEA WORM DISEASE

Dracunculasis or guinea worm disease is a preventable infection caused by the parasite *Dracunculus medinensis*.

Adult female Dracunculus worms emerge from the skin of infected persons annually. Persons with worms protruding through the skin may enter sources of drinking water and unwittingly allow the worm to release larvae into the water. These larvae are ingested by fresh water copepods where these develop into the infective stage in 10-14 days. Persons become infected by drinking water containing the water fleas harbouring the infective stage larvae of Dracunculus medinensis.

Infected people do not usually have symptoms until about 1 year after they become infected. It begins with a blister, usually on the leg. Around the time of its eruption, the person may experience itching, fever, swelling and burning sensations. Infected people try to relieve the pain by immersing the infected part in water, usually open water sources such as ponds and shallows wells. This allows the worm to spread.

At the beginning of the 20th century guinea worm disease was widespread in many countries in Africa and Asia. It is estimated that there were about 50 million cases in the 1950s. Due to concentrated efforts by international communities and endemic countries, the number of cases was reduced to about 96,000 by 1999.

FLUOROSIS

Fluorosis is an abnormal condition caused by excessive intake of fluorine, as from fluoridated drinking water, characterized chiefly by mottling of the teeth. Moderate level chronic exposure (above 1.5 mg/l of water) is quite common. Long-term ingestion of large amounts can lead to potentially severe skeletal problems (*skeletal fluorosis*).

The early symptoms of skeletal fluorosis include stiffness and pain in the joints. In severe case, the bone structure may change and ligaments may calcify, with resulting impairment of muscles and pain. Acute high-level exposure to fluoride causes immediate effects of abdominal pain, excessive saliva, nausea and vomiting. Seizures and muscle spasm may also occur.

It is believed that fluorosis affects millions of people around the world, but as regards dental fluorosis the very mild forms are the most frequent. Waters with high levels of fluoride content are mostly found at the foot of high mountains and in areas where the sea has made geological deposits. Known fluoride belt on land include: one that stretches from Syria through Jordan, Egypt, Libya, Algeria, Sudan and Kenya, and another that stretches from Turkey through Iraq, Iran, Afghanistan, India, northern Thailand and China.

GIARDIASIS

Giardiasis is an infection of the small intestine by a protozoan, *Giardia lamblia*. It is spread via the faecal-oral route, most commonly by eating food contaminated by the unwashed hands of an infected person or by drinking groundwater polluted by the faeces of infected animals.

About two thirds of infected individuals develop no symptoms. Symptoms, when present, occur one to three days after infection and consist of diarrhoea, flatulence, and abdominal cramps, often accompanied by weight loss. In some cases the infection becomes chronic. Chronic disease includes severe diarrhoea, gas in stomach, allergy, malnutrition, and stunted growth.

Giardiasis has traditionally been considered a tropical disease, but it is becoming more common in developed countries, especially among gay men and among groups of very young children in close contact with each other.

HEPATITIS

In medicine hepatitis is any disease featuring inflammation of the liver. Two of the viruses that cause hepatitis (hepatitis A and E) can be transmitted through water, food and from person to person. Hygiene is therefore important in their control. The illness starts with an abrupt onset of fever, body weakness, loss of appetite, nausea and abdominal discomfort, followed by jaundice within a few days.

The disease may range from mild (lasing 1-2 weeks) to severe disabling disease (lasting several months).

Both hepatitis A and E are found worldwide. Hepatitis A is particularly frequent in countries with poor sanitary and hygienic conditions (in Africa, Asia, and Central and South America). Countries with economies in transition and some regions of industrialized countries where sanitary conditions are sub-standard are also high affected (southern and eastern Europe and some parts of the Middle East).

HOOKWORM INFECTION

Hookworm infection is an intestinal parasite that usually causes diarrhoea or cramps. Heavy infestation with hookworm can be serious for newborns, children, pregnant women, and persons who are malnourished. Hookworm infection is contracted from contact with soil contaminated by hookworm, by walking barefoot or accidentally swallowing contaminated soil. The first signs of hookworm infection are itching and a rash at the site where the larvae penetrate the skin. These signs may be followed by abdominal pain, diarrhoea, loss of appetite and weight loss, and anaemia. Chronic heavy hookworm infection can damage the growth and development of children. The loss of iron and protein retards growth and mental development, sometimes irreversibly. Hookworm can also cause difficulty breathing, enlargement of the heart, and irregular heartbeat. Hookworm infections have been known to be fatal, particularly in infants.

They occur mainly in tropical and subtropical climates and affect about 1 billion people, about one-fifth of the world's population. One of the most common species of hookworm, *Ancylostoma duodenale,* is found in southern Europe, northern Africa, northern Asia, and parts of South America.

JAPANESE ENCEPHALITIS

Japanese encephalitis is a mosquito borne virus from the family *Flaviviridae.* Domestic pigs and wild birds are reservoirs of the virus; transmission to humans may cause severe symptoms. One of the most important Vectors of this disease is the mosquito *Culex tritaeniorhynchus.*

The disease is caused by a flavirus that affects the membranes around the brain.

Severe rigours mark the onset of this disease in Humans. Fever, Headache and malaise are other non-specific symptoms of this disease, which may last for a period between 1 to 6 days.

Signs, which develop during the acute encephalitic stage, include neck rigidity, cachexia, hemi paresis, convulsions and a raised body temperature between 38 and 41 degrees Celsius. Mental retardation developed from this disease usually leads to coma. Mortality of this disease varies but is generally much higher in children.

Japanese encephalitis is a leading cause of viral encephalitis in Asia with 30-50 million of clinical cases reported annually. Intensification and expansion of irrigated rice production systems in South-East Asia over the past 20 years have had an important burden caused by this disease. The flooding of the fields at the start of each cropping cycle leads to a big growth of the mosquito population.

LEAD POISONING

Lead poisoning is an intoxication of the system by organic compounds containing lead. These enter the body by respiration (of dust, fumes, or sprays) or by ingestion of food or other substances that contain lead. Water is rarely an important source of lead exposure except where lead pipes, for instance in old buildings, are common. Removal of old pipes is costly but the most effective measure to reduce lead exposure from water.

The symptoms of lead poisoning include reduced IQ, rashes, irritability, insomnia, excess lethargy or hyperactivity, poor appetite, headache. In high amounts it can result in abdominal discomfort, nervous system damage, and encephalitis.

At very high levels, lead can cause convulsions, coma and death. Chronic exposure is characterized by a blue line on the gums and can lead to damage to the brain, kidneys, nervous system, and red blood cells. The WHO guideline for lead in drinking water is 0.1 mg/l.

LEGIONELLOSIS

Legionellosis is an infection that is caused by the bacterium *Legionella pneumophila*. The disease has two distinct forms: legionnaires disease is the more severe form of the infection, which may involve pneumonia; Pontiac fever is a milder illness that develops from some hours to two days after initial infection and resolves spontaneously. Legionellosis infection occurs after persons have breathed mists that come from a water source (e.g., air conditioning cooling towers, whirlpool spas, showers) contaminated with Legionella bacteria.

People may be exposed to these mists in homes, workplaces, hospitals, or public places. Legionellosis is not passed from person to person, and there is no evidence of persons becoming infected from auto air conditioners or household window air-conditioning units.

Patients with Legionnaires disease usually have fever, chills, and a cough, which may be dry or may produce sputum. Some patients also have muscle aches, headache, tiredness, loss of appetite, and, occasionally, diarrhoea. Laboratory tests may show that these patients' kidneys are not functioning properly. Chest X-rays often show pneumonia. The bacteria reproduce to high numbers in warm, stagnant water (32°-40° C), such as that found in certain plumbing systems and hot water tanks, cooling towers and evaporative condensers of large air-conditioning systems, and whirlpool spas.

Legionnaires disease is believed to occur worldwide. It is estimated that approximately 10 to 15 thousand people contract it in the United States each year. An additional unknown number of people are infected with the Legionella bacterium but have only mild symptoms or no symptoms at all.

LEPTOSPIROSIS

Leptospirosis is an infectious disease of domestic animals, especially cattle, swine, and dogs, caused by spirochetes of the genus Leptospira and characterized by jaundice and fever. It is transmissible to humans through direct contact with the urine of infected animals or by contact with a urine-contaminated environment, such as surface water, soil and plants.

In humans it causes a wide range of symptoms, and some infected people may have no symptoms at all. Because of the wide range of symptoms the infection is often wrongly diagnosed. This leads to a lower registered number of cases than there really are.

Symptoms of leptospirosis include high fever, severe headache, chills, muscle aches, and vomiting, and may include jaundice, red eyes, abdominal pain, diarrhoea, or a rash. Complications include renal failure, meningitis, liver failure, and respiratory distress. In rare cases death occurs. Leptospirosis occurs worldwide, in both rural and urban areas and in temperate and tropical climates. The number of human cases worldwide is not well documented. It probably ranges from 0.1 to 1 per 100,000 per year in temperate climates to 10 or more per 100,000 may be infected. For several reasons leptospirosis is overlooked and consequently underreported in many areas of the world.

LYMPHATIC FILARIASIS

Lymphatic filarasis is a mosquito-borne parasitic worm infection. A person with the disease tends to have more bacterial infections in the skin and lymph system.

This causes hardening and thickening of the skin, which in its most dramatic form is expressed in the symptoms of elephantiasis, the accumulation of lymph, usually in legs. It is not a killer disease, but causes severe debilitation and social stigma.

The disease spreads from person to person by mosquito bites. Prevention includes giving entire communities medicine that kills the microscopic worms and controlling mosquitoes. Avoiding mosquito bites is another form of prevention. The mosquitoes that carry the microscopic worms usually bite between the hours of dusk and dawn.

MALARIA

Malaria is the world's most important parasitic disease transmitted from one person to another through the bite of female *Anopheles* mosquitoes, which breed in fresh or occasionally brackish water.

Its symptoms are at the onset of malaria, bouts of chills (ague) and fever lasting several hours and occurring every three or four days. If the disease is not treated, the spleen and the liver become enlarged, anaemia develops, and jaundice appears. General debility, anaemia, or clogging of the vessels of cerebral tissues by affected red blood cells can be followed by death.

Today, malaria occurs mostly in tropical and subtropical countries, particularly in Africa south of the Sahara, South-East Asia and the forest fringe zones in South America.

MALNUTRITION

Malnutrition is a general term for the medical condition in a person or animal caused by an unbalanced diet (either too little or too much food) or a diet missing one or more important nutrients. Clinically, malnutrition is characterized by inadequate or excess intake of proteins, energy and micronutrients such as vitamins, and the frequent infections and disorders result.

Malnutrition in all its forms increases the risk of disease and early death. Severe forms of malnutrition include marasmus, cretinism and irreversible brain damage due to iodine deficiency, blindness and increased risk of infection and death from vitamin A deficiency. Chronic food deficits affect about 792 million people in the world, including 20% of the population in developing countries.

Worldwide, malnutrition affects one in three people and each of its major forms influence most diseases globally. Malnutrition affects all age groups, but it is especially common among the poor and those with inadequate access to health education and clean water and good sanitation. More than 70% of children with malnutrition live in Asia, 26% in Africa, and 4% in Latin America and the Caribbean.

METHAEGLOBINEMIA

Methaeglobinemia is characterized by the reduced ability of the blood to carry oxygen because of reduced levels of normal haemoglobin. One of the most common causes is nitrate in drinking water. In the body nitrates are converted in nitrites. The nitrites react with the haemoglobin in the red blood cells to from methaemoglobin.

Controlling nitrate levels in drinking water sources to below around 50 mg/l is an effective preventive measure.

Is it uncommon, infants are the most affected, showing signs of blueness around the mouth, hands and feet, having trouble breathing as well as vomiting and diarrhoea. In extreme cases there is marked lethargy, an increase in the production of saliva, loss of consciousness, some cases can be fatal.

It is now rare in most of the industrialized countries due to control of nitrate contamination in water supplies. Occasional cases continue to be reported from rural areas. It is a risk in developing countries, for example where the drinking water is from shallow wells in farming areas.

ONCHOCERIASIS OR RIVER BLINDNESS

Onchoceriasis or river blindness is the world's second leading infectious cause of blindness. It is caused by *Onchocerca volvulus*, a parasitic worm, that breeds in water and that can live for up to fourteen years in the human body. Controlling insect breeding sites in rivers is one of the pillars of prevention. The disease is transmitted from one person to another through the bite of a black fly (*Simulium*).

Symptoms of the disease in a person usually begin to show 1-3 years after infection. Each adult female worm, which can be more than half a meter in length, produces millions of microscopic young worms (microfilaria). The microfilaria migrate through the skin and, upon death, cause intense itching and depigmentation of the skin (*leopard skin*), lymphadenitis resulting in hanging groins and elephantiasis of the genitals, serious visual impairment and blindness when they reach the eye.

Onchoceriasis is found in 36 countries in Africa as well as in Guatemala, southern Mexico, some areas of Venezuela, small areas in Brazil, Colombia and Ecuador, and in the Arabian peninsula. A total of 18 million people are affected worldwide.

POLIOMYELITIS (POLIO)

Polio (*infantile paralysis*) is a communicable disease, which is categorized as a disease of civilization. Polio spreads through human-to-human contact, usually entering the body through the mouth due to faecally contaminated water or food.

The disease is usually fatal if the nerve cells in the brain are attacked (*bulbar poliomyelitis*), causing paralysis of essential muscles, such as those controlling swallowing, heartbeat, and respiration.

There is no specific drug for treatment. For reasons not clearly understood, some people who have had severe polio experience post polio syndrome, a condition in which new weakness and pain occurs years later in previously affected muscles.

Polio cases have decreased by over 99% since 1988, from an estimated more than 350,000 cases to 1919 in 2003. The reduction is the result of the global effort to eradicate the disease. The seven polio-endemic countries/areas, from highest to lowest risk of ongoing transmission beyond mid-2003 are northern India, northern Nigeria, Egypt, Pakistan, Afghanistan, Somalia and Niger.

RING WORM OR TINEA

Ring worm or tinea is superficial eruption of the skin caused by a fungus, chiefly *Microsporum*, *Trichophyton*, or *Epidermophyton*. The most common affected area is the feet, but any area of the skin may be affected, including the scalp and nails.

Fungi are present on the bodies of most persons, but some individuals are more resistant to fungus invasion than others.

It is spread by direct contact with an infected person or animal, contact with soil or by indirect contact with items contaminated by the fungus (clothing, towels, bedclothes and toilet articles). On the scalp, ringworm begins in the form of a pimple sore, which then spreads into a ring shape. Hair becomes brittle, breaking easily and falling out, leaving bald spots on the scalp. On the body, ringworm may first appear as red or pink, flat or slightly raised, patched on the skin. Infection in the nails usually begins at the site of an injured nail and may spread to the other nails. Infected nails become thick, pitted, grooved and abnormal in shape and colour. Although specific figures are not available, ringworm is a frequent problem in most countries, particularly where personal and domestic hygiene are poor.

SCABIES

Scabies is a highly contagious parasitic skin disease caused by the itch mite (*Sarcoptes scabiei*). It is acquired through close contact with an infested individual or contaminated clothing and is most prevalent among those living in crowded and unhygienic conditions. Personal hygiene is an important preventive measure and access to adequate water supply is important in control.

The principal sign of the disease is a pimple-like rash that is most commonly found on hands, especially the webbing between the fingers, the skin folds of the wrist, elbow or knee, the penis, the breast or the shoulder. Infestation often causes intense itching all over the body, especially at night.

Scabies mites are found worldwide, affecting all socioeconomic classes and in all climates. Epidemics have been linked to poverty, poor water supply, sanitation, and overcrowding. There are about 300 million cases of scabies in the world each year.

ESQUISTOMIASIS

La esquistomiasis es una enfermedad parasitaria causada en periodos de inundaciones, sitic disease caused by blood flukes, trematode worms of the genus *Schistosoma*. The larvae of the parasite are harboured by snails and infect humans who bathe in or come in contact with infested waters. The larvae enter through the skin, migrate via the blood vessels, and mature in the lungs. From there they travel to the veins of the upper or lower intestine or bladder and reproduce.

The signs following the infection are rashes or itchy skin. Two months after infection, fever, chills, cough and muscle aches may occur, as the parasites mature. Untreated infections can result in blood in urine and stools, and enlarged liver and spleen.

Other symptoms depend on the tissues affected; cirrhosis of the liver is common. Although symptoms vary according to the species of infecting fluke, all forms can result in general weakening and eventual death.

Schistosomiasis is considered the second most important parasitic infection after malaria in terms of public health and economic impact. The disease is prevalent in Asia, some Pacific islands, Africa, the West Indies, South America, Spain, Puerto Rico, and Cyprus.

TRACHOMA

Trachoma is an infection of the mucous membrane of the eyelids caused by the bacterium *Chlamydia trachomatis.*

It begins as congestion and swelling of the eyelids with tearing and disturbance of vision. The cornea is often involved. If left untreated, scar tissue forms, which causes deformities of the eyelids and, if there is corneal involvement, partial or total blindness.

It begins as congestion and swelling of the eyelids with tearing and disturbance of vision. The cornea is often involved. If left untreated, scar tissue forms, which causes deformities of the eyelids and, if there is corneal involvement, partial or total blindness.

Trachoma is highly contagious in its early stages and is transmitted by direct contact with infected persons or articles (e.g., towels, handkerchiefs) and possibly also by flies.

Trachoma is the world leading cause of preventable blindness and occurs where people live in overcrowded conditions with the limited access to water and health care. Blinding trachoma is widespread in the Middle East, North and Sub Sahara Africa part of the Indian subcontinent, Southern Asia and China. The WHO estimates that 6 million worldwide are blind due to trachoma and more than 150 million people are in need of treatment.

TRICHURIASIS

Trichuriasis is a parasitic disease cause by infection of the large intestine by a parasite whipworm (Trichuris trichiura). The main risk factor for infection is ingestion of eggs from soil contaminated with faeces. Some outbreaks have been traced to contaminated vegetables (due to presumed soil contamination).

The development of symptoms depends on the burden of worms/eggs. Light infestations are frequently asympto-matic, heavy infestations may have bloody diarrhoea, long-standing blood loss may lead to iron-deficiency anaemia, rectal prolepses is seen in severe cases.

Trichuriasis is common worldwide, in particular among countries with warm, humid climates, and primarily affects children.

TYPHOID AND PARATYPHOID ENTERIC FEVER

Typhoid and paratyphoid enteric fever are acute, generalized infections caused by Salmonella typhi and Salmonella paratyphoid respectively. The main sources of infection are contaminated water or milk and, especially in urban communities, food handlers who are carriers. Their germs are passed in the faeces and urine of infected people. People become infected after eating food or drinking beverages that have been handled by a person who is infected or by drinking water that has been contaminated by sewage containing the bacteria.

Once the bacteria enter the person's body they multiply and spread from the intestines, into the bloodstream. The symptoms of typhoid appear 10 to 14 days after infection; they can be mild or severe and include high fever, rose-colored spots on the abdomen and chest, diarrhoea or constipation, and enlargement of the spleen and liver. In untreated patients complications may be numerous, affecting practically every body system, and can even include perforation of the intestine with haemorrhage.

Complications account for the mortality rate of 7% to 14%. The annual incidence of typhoid is estimated to be about 17 million cases worldwide. Typhoid and paratyphoid fevers are common in less-industrialized countries, principally owing to they problem of unsafe drinking water, inadequate sewage disposal and flooding.

Chapter 7

Nitrogen Cycle

The nitrogen cycle is the process by which nitrogen is converted between its various chemical forms. This transformation can be carried out via both biological and non-biological processes. Key processes in the nitrogen cycle include fixation, mineralization, nitrification, and denitrification.

The majority of Earth's atmosphere (approximately 78-80%) is nitrogen, making it the largest pool of nitrogen. However, atmospheric nitrogen is unavailable for biological use, leading to a scarcity of usable nitrogen in many types of ecosystems.

The nitrogen cycle is of particular interest to ecologists because nitrogen availability can affect the rate of key ecosystem processes, including primary production and decomposition. Human activities such as fossil fuel combustion, use of artificial nitrogen fertilizers, and release of nitrogen in wastewater have dramatically altered the global nitrogen cycle.

ECOLOGICAL FUNCTION

Nitrogen is essential for many processes; it is crucial for any life on Earth. It is in all amino acids, is incorporated into proteins, and is present in the bases that make up nucleic acids, such as DNA and RNA. In plants, much of the nitrogen is used in chlorophyll molecules, which are essential for photosynthesis and further growth.

Although earth's atmosphere is an abundant source of nitrogen, most is relatively unusable by plants. Chemical processing, or natural fixation (through processes such as bacterial conversion—see rhizobium), are necessary to convert gaseous nitrogen into forms usable by living organisms. This makes nitrogen a crucial part of food production. The abundance or scarcity of this "fixed" form of nitrogen, (also known as reactive nitrogen), dictates how much food can be grown on a piece of land.

ENVIRONMENTAL IMPACTS

Increased use of nitrogen fertilizers is causing dramatic environmental changes, including surface and groundwater pollution, ocean dead zones and boosting global warming.

Ammonia is highly toxic to fish and the water discharge level of ammonia from wastewater treatment plants must often be closely monitored. To prevent loss of fish, nitrification prior to discharge is often desirable. Land application can be an attractive alternative to the mechanical [aeration] needed for nitrification.

The Processes of the Nitrogen Cycle

Nitrogen is present in the environment in a wide variety of chemical forms including organic nitrogen, ammonium (NH_4^+), nitrate (NO_3^-), and nitrogen gas (N_2). The processes of the nitrogen cycle transform nitrogen from one chemical form to another. Many of the processes are carried out by microbes either to produce energy or to accumulate nitrogen in the form needed for growth.

NITROGEN FIXATION

Nitrogen fixation usually refers to the biological process by which nitrogen (N_2) in the atmosphere is converted into ammonia. This process is essential for life because fixed nitrogen is required to biosynthesize the basic building blocks of life, e.g. nucleotides for DNA and amino acids for proteins. Formally, nitrogen fixation also refers to other abiological conversions of nitrogen, such as its conversion to nitrogen dioxide.

Nitrogen fixation is utilized by numerous prokaryotes,

including bacteria, actinobacteria, and certain types of anaerobic bacteria. Microorganisms that fix nitrogen are called diazotrophs. Some higher plants, and some animals (termites), have formed associations (symbioses) with diazotrophs. Nitrogen fixation also occurs as a result of non-biological processes. These include lightning, industrially through the Haber-Bosch Process, and combustion. Biological nitrogen fixation was discovered by the Dutch microbiologist Martinus Beijerinck.

Biological Nitrogen Fixation

Biological nitrogen fixation (BNF) occurs when atmospheric nitrogen is converted to ammonia by an enzyme called nitrogenase.

The formula for BNF is:

$$N_2 + 6\,H^+ + 6\,e^- \rightarrow 2\,NH_3$$

The process is coupled to the hydrolysis of 16 equivalents of ATP and is accompanied by the co-formation of one molecule of H_2. In free-living diazotrophs, the nitrogenase-generated ammonium is assimilated into glutamate through the glutamine synthetase/glutamate synthase pathway.

Enzymes responsible for nitrogenase action are very susceptible to destruction by oxygen. (In fact, many bacteria cease production of the enzyme in the presence of oxygen). Many nitrogen-fixing organisms exist only in anaerobic conditions, respiring to draw down oxygen levels, or binding the oxygen with a protein such as Leghemoglobin.

Plants that contribute to nitrogen fixation include the legume family – Fabaceae – with taxa such as clover, soybeans, alfalfa, lupines and peanuts. They contain symbiotic bacteria called *Rhizobia* within nodules in their root systems, producing nitrogen compounds that help the plant to grow and compete with other plants.

When the plant dies, the fixed nitrogen is released, making it available to other plants and this helps to fertilize the soil The great majority of legumes have this association, but a few genera (e.g., *Styphnolobium*) do not. In many traditional and organic farming practices, fields are rotated through various types of crops, which usually includes one consisting mainly or entirely of clover or buckwheat (family *Polygonaceae*), which were often referred to as "green manure."

Microorganisms that fix nitrogen:

- Diazotrophs
- Cyanobacteria
- *Azotobacteraceae*
- *Rhizobia*
- *Frankia*

Nitrogen Fixation by Cyanobacteria

Cyanobacteria inhabit nearly all illuminated environments on Earth and play key roles in the carbon and nitrogen cycle of the biosphere. Generally, cyanobacteria are able to utilize a variety of inorganic and organic sources of combined nitrogen, like nitrate, nitrite, ammonium, urea or some amino acids.

Several cyanobacterial strains are also capable of diazotrophic growth. Genome sequencing has provided a large amount of information on the genetic basis of nitrogen metabolism and its control in different cyanobacteria. Comparative genomics, together with functional studies, has led to a significant advance in this field over the past years. 2-oxoglutarate has turned out to be the central signalling molecule reflecting the carbon/nitrogen balance of cyanobacteria.

Central players of nitrogen control are the global transcriptional factor NtcA, which controls the expression of many genes involved in nitrogen metabolism, as well as the P_{II} signalling protein, which fine-tunes cellular activities in response to changing C/N conditions.

These two proteins are sensors of the cellular 2-oxoglutarate level and have been conserved in all cyanobacteria. In contrast, the adaptation to nitrogen starvation involves heterogeneous responses in different strains. Nitrogen fixation by cyanobacteria in coral reefs can fix twice the amount of nitrogen than on land – around 1.8kg of nitrogen is fixed per hectare per day.

Chemical Nitrogen Fixation

Nitrogen can also be artificially fixed. for use in fertilizers, explosives, or in other products. The most common method is the Haber process. Artificial fertilizer production is now the largest source of human-produced fixed nitrogen in the Earth's ecosystem.

The Haber process requires high pressures (around 200 atm) and high temperatures (at least 400°C), routine conditions for industrial catalysis. This highly efficient process uses natural gas as a hydrogen source and air as a nitrogen source.

Research on Catalytic Nitrogen-fixation

Much research has been conducted on the discovery of catalysts for nitrogen fixation, often with the goal of reducing the energy required for this conversion. Such research has thus far failed to even approach the efficiency and ease of the Haber process, however. Many compounds react with atmospheric nitrogen under ambient conditions. For example, lithium metal converts to lithium nitride under an atmosphere of nitrogen. Treatment of the resulting nitride gives ammonia.

The first dinitrogen complex was reported in 1965 based on ammonia coordinated to ruthenium ($[Ru(NH_3)5(N_2)]^{2+}$), research in chemical fixation focused on transition metal complexes. Since then a large number of transition metal compounds that contain dinitrogen as a ligand have been discovered. The dinitrogen ligand can be bound either to a single metal or bridge two (or more) metals. The coordination chemistry of dinitrogen is complex and currently under intense investigation. This research may lead to new ways of using dinitrogen in synthesis and on an industrial scale.

The first example of homolytic cleavage of dinitrogen under mild conditions was published in 1995. Two equivalents of a molybdenum complex reacted with one equivalent of dinitrogen, creating a triple bonded MoN complex. Since then, this triple bounded complex has been used to make nitriles. The first catalytic system converting nitrogen to ammonia at room temperature and pressure was discovered in 2003 and is based on another molybdenum compound, a proton source and a strong reducing agent. Unfortunately, this catalytic reduction only fixates a few nitrogen molecules.

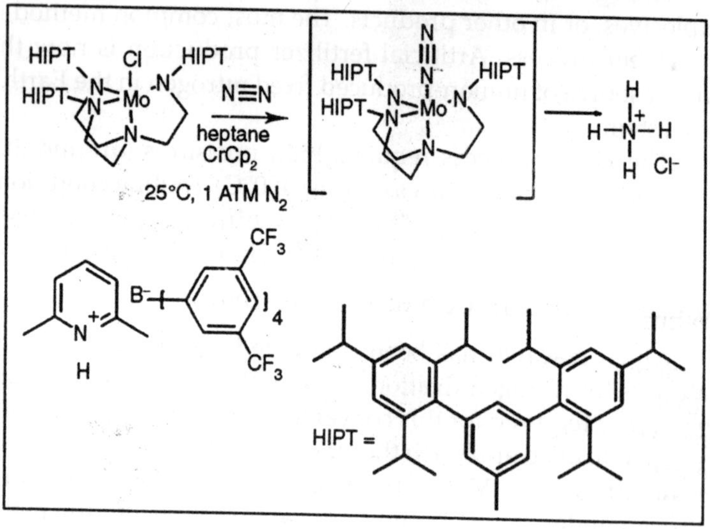

CONVERSION OF N_2

The conversion of nitrogen (N_2) from the atmosphere into a form readily available to plants and hence to animals and humans is an important step in the nitrogen cycle, which distributes the supply of this essential nutrient.

There are four ways to convert N_2 (atmospheric nitrogen gas) into more chemically reactive forms:

- Biological fixation: some symbiotic bacteria (most often associated with leguminous plants) and some free-living bacteria are able to fix nitrogen as organic nitrogen. An example of mutualistic nitrogen fixing bacteria are the *Rhizobium* bacteria, which live in legume root nodules. These species are diazotrophs. An example of the free-living bacteria is *Azotobacter*.
- Industrial N-fixation: Under great pressure, at a temperature of 600° C, and with the use of an iron catalyst, atmospheric nitrogen and hydrogen (usually derived from natural gas or petroleum) can be combined to form ammonia (NH_3). In the Haber-

Bosch process, N_2 is converted together with hydrogen gas (H_2) into ammonia (NH_3), which is used to make fertilizer and explosives.

- Combustion of fossil fuels: automobile engines and thermal power plants, which release various nitrogen oxides (NO_x).
- Other processes: In addition, the formation of NO from N_2 and O_2 due to photons and especially lightning, are important for atmospheric chemistry, but not for terrestrial or aquatic nitrogen turnover.

Assimilation

Some plants get nitrogen from the soil, and by absorption of their roots in the form of either nitrate ions or ammonium ions. All nitrogen obtained by animals can be traced back to the eating of plants at some stage of the food chain.

Plants can absorb nitrate or ammonium ions from the soil via their root hairs. If nitrate is absorbed, it is first reduced to nitrite ions and then ammonium ions for incorporation into amino acids, intense nucleic acids, and chlorophyll. In plants that have a mutualistic relationship with rhizobia, some nitrogen is assimilated in the form of ammonium ions directly from the nodules.

Animals, fungi, and other heterotrophic organisms absorb nitrogen as amino acids, nucleotides and other small organic molecules.

Ammonification

When a plant dies, an animal dies, or an animal expels waste, the initial form of nitrogen is organic. Bacteria, or in some cases, fungi, convert the organic nitrogen within the remains back into ammonium (NH_4^+), a process called ammonification or mineralization.

Enzymes Involved:

- *GS*: Gln Synthetase (Cytosolic & PLastid)
- *GOGAT*: Glu 2-oxoglutarate aminotransferase (Ferredoxin & NADH dependent)
- *GDH*: Glu Dehydrogenase:

- Minor Role in ammonium assimilation.
- Important in amino acid catabolism.

Nitrification

Nitrification is the biological oxidation of ammonia with oxygen into nitrite followed by the oxidation of these nitrites into nitrates. Degradation of ammonia to nitrite is usually the rate limiting step of nitrification. Nitrification is an important step in the nitrogen cycle in soil. This process was discovered by the Russian microbiologist, Sergei Winogradsky.

Microbiology and Ecology

The oxidation of ammonium into nitrite is performed by two groups of organisms, ammonia oxidizing bacteria and ammonia oxidizing archaea. Ammonia oxidizing bacteria can be found among the â- and ã-proteobacteria. In soils the most studied ammonia oxidizing bacteria belong to the genera Nitrosomonas and Nitrosococcus.

Although in soils ammonia oxidation occurs by both bacteria and archaea, archaeal ammonia oxidizers dominate in both soils and marine environments, suggesting that Crenarchaeota may be greater contributors to ammonia oxidation in these environments.

The second step (oxidation of nitrite into nitrate) is (mainly) done by bacteria of the genus Nitrobacter. Both steps are producing energy to be coupled to ATP synthesis. Nitrifying organisms are chemoautotrophs, and use carbon dioxide as their carbon source for growth.

Nitrification also plays an important role in the removal of nitrogen from municipal wastewater. The conventional removal is nitrification, followed by denitrification. The cost of this process resides mainly in aeration (bringing oxygen in the reactor) and the addition of an external carbon source (e.g. methanol) for the denitrification.

In most environments both organisms are found together, yielding nitrate as the final product. It is possible however to design systems in which selectively nitrite is formed (the Sharon process).

Together with ammonification, nitrification forms a mineralization process which refers to the complete decomposition of organic material, with the release of available nitrogen compounds. This replenishes the nitrogen cycle.

Chemistry

Nitrification is a process of nitrogen compound oxidation (effectively, loss of electrons from the nitrogen atom to the oxygen atoms):

- $NH_3 + CO_2 + 1.5\ O_2$ + Nitrosomonas $\rightarrow NO_2^- + H_2O + H^+$
- $NO_2^- + CO_2 + 0.5\ O_2$ + Nitrobacter $\rightarrow NO_3^-$
- $NH_3 + O_2 \rightarrow NO_2 + 3H^+ + 2e$
- $NO_2 + H_2O \rightarrow NO_3 + 2H^+ + 2e$

DENITRIFICATION

Denitrification is a microbially facilitated process of dissimilatory nitrate reduction that may ultimately produce molecular nitrogen (N_2) through a series of intermediate gaseous nitrogen oxide products.

This respiratory process reduces oxidized forms of nitrogen in response to the oxidation of an electron donor such as organic matter. The preferred nitrogen electron acceptors in order of most to least thermodynamically favourable include: nitrate (NO_3^-), nitrite (NO_2^-), nitric oxide (NO), and nitrous oxide (N_2O).

In terms of the general nitrogen cycle, denitrification completes the cycle by returning N_2 to the atmosphere. The process is performed primarily by heterotrophic bacteria (such as *Paracoccus denitrificans* and various pseudomonads), although autotrophic denitrifiers have also been identified (e.g.,*Thiobacillus denitrificans*).

Denitrifiers are represented in all main phylogenetic groups. Generally several species of bacteria are involved in the complete reduction of nitrate to molecular nitrogen, and more than one enzymatic pathway have been identified in the reduction process. Denitrification takes place under special conditions in both terrestrial and marine ecosystems. In general, it occurs where oxygen, a more energetically favourable electron acceptor, is depleted, and bacteria respire nitrate as a substitute terminal electron acceptor.

Due to the high concentration of oxygen in our atmosphere, denitrification only takes place in environments where oxygen consumption exceeds the rate of oxygen supply, such as in some soils and groundwater, wetlands, poorly ventilated corners of the ocean, and in seafloor sediments.

Denitrification generally proceeds through some combination of the following intermediate forms:

$$NO_3 \rightarrow NO_2 \rightarrow NO + N_2O \rightarrow N_2\ (g)$$

The complete denitrification process can be expressed as a redox reaction:

$$2\ NO_3 + 10\ e + 12\ H^+ \rightarrow N_2 + 6\ H_2O$$

This reaction shows a fractionation in isotope composition. Lighter isotopes of nitrogen are preferred in the reaction, leaving the heavier nitrogen isotopes in the residual matter. The process can cause delta-values of up to -40, where delta is a representation of the difference in isotopic composition. This can be used to identify denitrification processes in nature.

Denitrification is commonly used to remove nitrogen from sewage and municipal wastewater. It is also an instrumental process in wetlands and riparian zones for the removal of excess nitrate from groundwater resulting from excessive agricultural or residential fertiliser usage. Direct reduction from nitrate to ammonium, a process known as dissimilatory nitrate reduction to ammonium or DNRA, is also possible for organisms that have the nrf-gene. This is less common than denitrification in most ecosystems as a means of nitrate reduction. *Reduction under anoxic conditions can also occur through process called anaerobic ammonia oxidation (anammox):*

$$NH_4^+ + NO_2^{''} \rightarrow N_2 + 2\ H_2O$$

In some wastewater treatment plants, small amounts of methanol are added to the wastewater to provide a carbon source for the denitrification bacteria.

Anaerobic Ammonium Oxidation

In this biological process, nitrite and ammonium are

converted directly into dinitrogen gas. This process makes up a major proportion of dinitrogen conversion in the oceans.

Human Influences on the Nitrogen Cycle

As a result of extensive cultivation of legumes (particularly soy, alfalfa, and clover), growing use of the Haber-Bosch process in the creation of chemical fertilizers, and pollution emitted by vehicles and industrial plants, human beings have more than doubled the annual transfer of nitrogen into biologically available forms. In addition, humans have significantly contributed to the transfer of nitrogen trace gases from Earth to the atmosphere, and from the land to aquatic systems.

Human alterations to the global nitrogen cycle are most intense in developed countries and in Asia, where vehicle emissions and industrial agriculture are highest.

N_2O (nitrous oxide) has risen in the atmosphere as a result of agricultural fertilization, biomass burning, cattle and feedlots, and other industrial sources.

N_2O has deleterious effects in the stratosphere, where it breaks down and acts as a catalyst in the destruction of atmospheric ozone.

N_2O in the atmosphere is a greenhouse gas, currently the third largest contributor to global warming, after carbon dioxide and methane. While not as abundant in the atmosphere as carbon dioxide, for an equivalent mass, nitrous oxide is nearly 300 times more potent in its ability to warm the planet.

Ammonia (NH_3) in the atmosphere has tripled as the result of human activities. It is a reactant in the atmosphere, where it acts as an aerosol, decreasing air quality and clinging on to water droplets, eventually resulting in acid rain. Fossil fuel combustion has contributed to a 6 or 7 fold increase in NOx flux to the atmosphere.

NO_2 actively alters atmospheric chemistry, and is a precursor of tropospheric (lower atmosphere) ozone production, which contributes to smog, acid rain, damages plants and increases nitrogen inputs to ecosystems. Ecosystem processes can increase with nitrogen fertilization, but anthropogenic input can also result in nitrogen saturation, which weakens productivity and can kill plants. Decreases in biodiversity can also result if higher nitrogen availability increases nitrogen-demanding grasses, causing a degradation of nitrogen-poor, species diverse heathlands.

Wastewater Treatment

Onsite sewage facilities such as septic tanks and holding tanks release large amounts of nitrogen into the environment by discharging through a drainfield into the ground. Microbial activity consumes the nitrogen and other contaminants in the wastewater. However, in certain areas, the soil is unsuitable to handle some or all of the wastewater, and, as a result, the wastewater with the contaminants enters the aquifers. These contaminants accumulate and eventually end up in drinking water. One of the contaminants concerned about the most is nitrogen in the form of nitrates. A nitrate concentration of 10 ppm or 10 milligrams per liter is the current EPA limit for drinking water and typical household wastewater can produce a range of 20-85 ppm.

The health risk associated with drinking water (with >10 ppm nitrate) is the development of methemoglobinemia and has been found to cause blue baby syndrome. Several states have now started programs to introduce advanced wastewater treatment systems to the typical onsite sewage facilities. The result of these systems is an overall reduction of nitrogen, as well as other contaminants in the wastewater.

Environmental Damage

Additional risks posed by increases in fixed nitrogen in aquatic systems include spurring the creation and growth of eutrophic lakes and oceanic dead zones through algal bloom-induced hypoxia.

The extent and effects of the anthropogenically-induced doubling of biologically available nitrogen in the soils, waters, and air of the earth during the past century are still poorly understood.

THE BIOGEOCHEMICAL NITROGEN CYCLE

The biogeochemical nitrogen cycle is complex with nitrogen occurring in valence states from -3 to +5. Although abundant on earth, 96 per cent is found in the lithosphere and does not take part in the biogeochemical cycle.

In terrestrial systems, only four per cent of the nitrogen occurs in biomass, while 96 per cent is found mainly in soil organic matter.

Nitrogen is an essential nutrient but nitrogen compounds are also potential toxicants. The hazard of unwanted side effects is increasing, i.e. through the increased use of nitrogen fertilizers. By the end of this century, man made additions of combined nitrogen to terrestrial ecosystems will be as large as the amount added through biological nitrogen fixation.

An understanding of the individual processes of the nitrogen biogeochemical cycle is needed before attempts can be made to quantify the nitrogen cycle of ecosystems, regions, or the earth. The inorganic nitrogen cycle, which was earlier considered relatively simple, is becoming more complex as our knowledge increases.

To establish a quantitative nitrogen budget for a system, we need extrapolation in time and space. Extrapolation in space is difficult due to the patchy distribution of anaerobic microsites. Extrapolation in time calls for a thorough understanding of the factors affecting process rates in order to quantify yearly rates based on a few measurements only.

The nitrogen cycle in an ecosystem is an important characteristic. The pattern of nitrogen cycling changes as an ecosystem develops to its climax state, and it is also affected by disturbances such as management practices.

A quantification and understanding of the nitrogen cycle in various types of typical ecosystem of the world will offer us the jigsaw pieces needed to construct an authoritative picture of the global nitrogen cycle, intrinsically linked to other major elements such as carbon, phosphorus and sulphur.

Few elements are as complex and interesting as nitrogen. Firstly, this complexity is reflected in the highly intricate biogeochemicalcycle, where nitrogen occurs in valence states from -3 to +5 and where many of the transformations are carried out by a few organisms only, at normal temperatures and pressures.

Table. Distribution of Nitrogen on Earth. All Values in Tg (10^{12}g). Data from Sweeney et.al

	Amount of Nitrogen (Tg)	Proportion (%)
Lithosphere	$574 \times 10^8$93.8	
Atmosphere	38×10^8	6.2
Hydrosphere	0.23×10^8	0.04
Biosphere	0.009×10^8	0.001

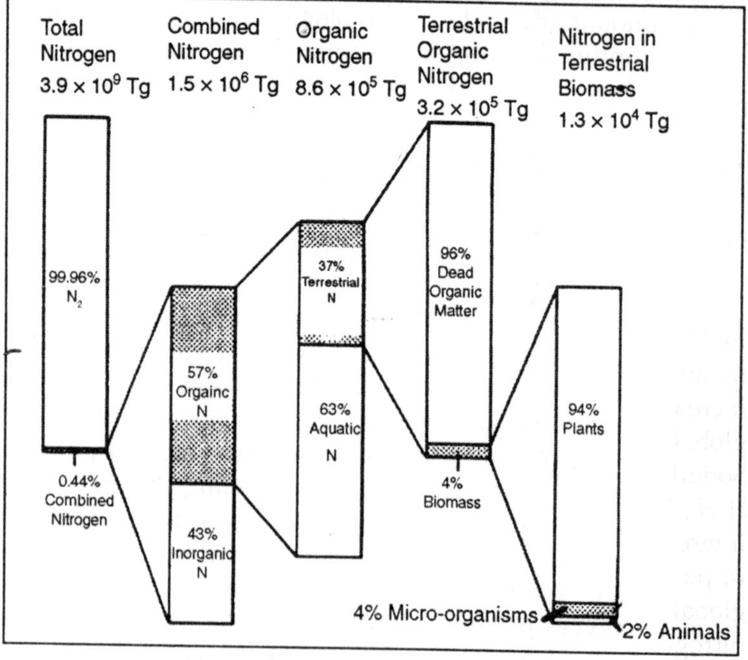

Fig. Distribution of Nitrogen in the Biosphere

Secondly, nitrogen is an element which is abundant on earth, but only a very small proportion of it enters into the biogeochemical nitrogen cycle at significant rates. Of the nitrogen found on earth, only 0.001 per cent occurs in the biosphere. Most of the nitrogen is thus found in the lithosphere (especially in primary rocks of the mantle). The atmosphere, where nitrogen occurs abundantly in its molecular form, still only contains six per cent of all nitrogen on earth.

Of all nitrogen taking part in the biogeochemical nitrogen cycle only 0.04 per cent occurs in compounds potentially available to living organisms. In the terrestrial system only four per cent occurs in biomass, the remainder forming a large reservoir as dead organic matter (96 per cent).

The unavailability of the 99.96 per cent occurring as nitrogen gas, combined with the major role played by nitrogen-containing substances in all forms of life, has caused nitrogen to be one of the key elements limiting the primary production on which man depends for his supply of food, fodder, fibre, and fuel. Thirdly, the nitrogen cycle is easily manipulated by man, and it has been estimated that, by the end of this century, man-made additions of combined nitrogen will equal the amounts fixed annually through biological nitrogen fixation. This increased addition from fertilizers, together with nitrogen oxides emitted into the atmosphere as a result of combustion, is undesirable since nitrogen compounds are a direct environmental hazard.

A SCOPE project on the global nitrogen cycle was started in 1974. It was terminated by the publication of a report quantifying the global cycle in detail not earlier attempted. The increased interest in the nitrogen cycle is reflected in the many global nitrogen budgets that have recently been published, e.g. Soderlund and Svensson, Delwiche, Hahn and Junge, Sweeney *et al.*, NAS, and Bolin. One of the main reasons behind this interest was the concern that nitrogen oxides may be acting as important catalytic agents regulating the thickness of the global stratospheric ozone shield. Our understanding of the nitrogen cycle is still, however, very rudimentary, and only by trying to understand the main factors regulating the many processes in the biogeochemical nitrogen cycle will it be possible to attain an improved quantitative picture of the global nitrogen cycle. An authoritative quantitative description is needed before we can quantify man's interventions in this cycle and their possible deleterious effects on the environment.

NITROGENMETABOLISMIN ECOLOGICALTERMS

Nitrogen Assimilation

Plants and most microorganisms are dependent on ammonium or nitrate salts for growth, though they seem generally to prefer ammonium as a nitrogen source. Ammonium nitrogen can be metabolized by two different pathways-either through glutamate dehydrogenase(GDH)or through glutamine synthetase glutamate synthase (GS/ GOGAT). Although GDH was previously considered the major enzyme involved in ammonium assimilation, it now seems clear that, especially at low ammonium concentrations, GS/ GOGATis the major enzyme system involved both in microorganisms and in plants. This latter system has a much higher affmity for the substrate (*K*m for GS: 1-2 × 10^{-5} M) than glutamate dehydrogenase (*Km* =4 X 10^{-3} M).It should be noted that glutamine synthetase is ATP-dependent, whereas glutamate dehydrogenase is not. This system is thus probably the rate-limiting step, since ammonia uptake seems to be a passive process.

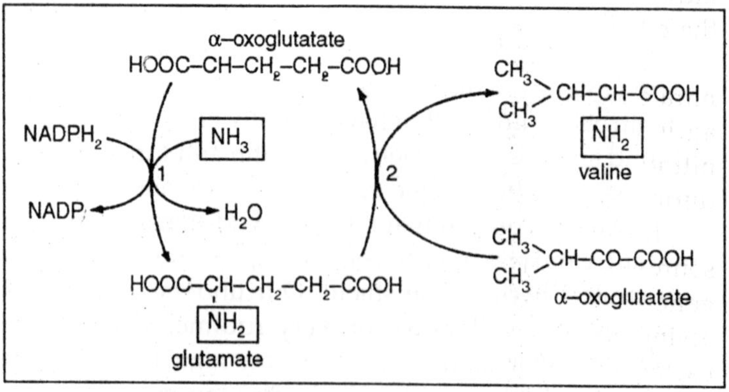

Fig. Assimilation of Ammonia (a) by Glutamate Dehydrogenase (1) and Subsequent Transfer of the Amino Group by a Transaminase (2)

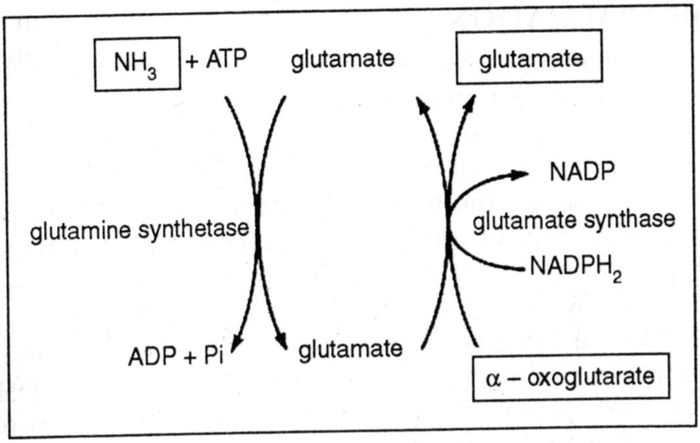

Fig. By Glutamine Synthetase/Glutamate Synthase (GS/GOGAT).

All plants, except certain bog species, are able to utilize nitrate as well as ammonium. If no other factors limit microbial growth, microorganisms will efficiently Scavenge the surroundings for available ammonium nitrogen. Since ammonium oxidation through nitrification proceeds rapidly in most soils, nitrate is probably the most important nitrogen source for most plants.

In contrast to ammonium uptake, plant root uptake of nitrate is dependent on a permease system. The occurrence of nitrate reductase in the plant seems to give an indication of the rate of nitrate supply in the plant.

A comparison of net assimilation or net growth with nitrate reductase activity at different points in time could give an indication of the relative importance of ammonium and nitrate as the nitrogen source for plants at different times during the growing season.

Microorganisms generally prefer ammonium as a nitrogen source, and their ability to use nitrate is restricted. Of the 2500 genera of fungi described, only 20 have been reported to assimilate nitrate. The occurrence of nitrate assimilation in bacteria seems to be more common than in fungi, although it is in no way ubiquitous. Since the assimilatory nitrate reductase is repressed by ammonium, the latter is the preferential nitrogen source for microorganisms.

MINERALIZATION

Ammonium is liberated through mineralization, mainly by microorganisms, from organic compounds. Inunobilization is the opposite process, where by inorganic nitrogen is assimilated and built up into organic compounds. For soil microorganisms, the balance between mineralization and immobilization of ammonium nitrogen, i.e. positive or negative nitrogen net mineralization, is regulated primarily by the C/N ratio of the substrate.

Microbial biomass in soil has been estimated to contain four per cent nitrogen, and if the carbon content is 50 per cent, the microbial biomass has a C/N ratio of 12.5. When an organic substrate is broken down by microorganisms, the quality of the substrate will determine the relative proportion of carbon assimilated and respired. If the assimilatory efficiency is 40 per cent, no net mineralization will occur if the C/N ratio is above 31. There are, however, very few data available on the assimilatory efficiency of microorganisms growing in soil, and this efficiency is very critical in determining the C/N ratio below which net mineralization occurs. However, one cannot judge the mineralization capacities of soils from their C/N ratios alone. In systems which accumulated organic matter, the C/N ratio is generally about 30. In tundra peat it has been observed to be as high as 48 in the top 10 cm, where most of the net mineralization occurs. Scots pine needle litter from a coniferous forest on glacial sediment in central Sweden originally had a C/N ratio as high as 120, whereas at the time when net mineralization started it was 80. If microbial biomass contains four per cent N, carbon/nitrogen ratios of 48 and 120 would require assimilatory efficiencies as low as 23 and 10 per cent respectively for net mineralization to occur.

One of the main difficulties in using such a crude concept as the C/N ratio of the substrate for determining whether mineralization or immobilization occurs relates to the forms of nitrogen in organic matter. Part of the nitrogen, especially which bound in proteins, is often easily mineralized, but another part, bound to the lignin fraction, is very resistant to mineralization. Such differences must be kept in mind in any attempt to describe net nitrogen mineralization. Soil animals may play an important role in regulating nitrogen mineralization in at least two ways.

The effect of faunal grazing on microorganisms may increase mineralization rates. It also seems that soil invertebrates may play an important role through their excretion of significant amounts of simple nitrogenous substances, such as uric acids, urea, and also ammonia. They may thus playa more important role in nitrogen mineralization than hitherto assumed.

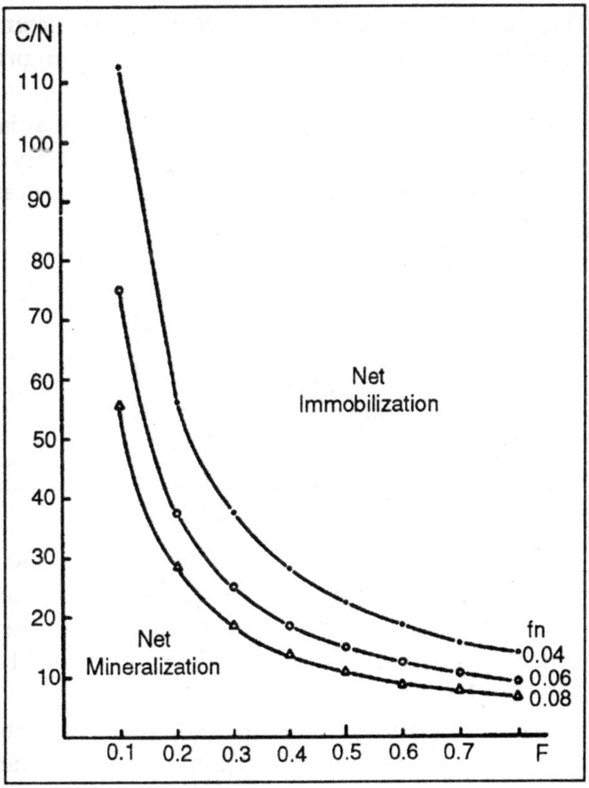

Fig. The Dependence of Net Mineralization Versus Immobilization on the Carbon/Nitrogen Ratio of the Substrate, the Assimilatory Efficiency of the Decomposers *(F)*, and the Nitrogen Content of Microbial Biomass *(fn)*

Nitrification-Denitrification

Nitrogen also plays an important role in generating energy in certain groups of microorganisms. Autotrophic nitrifying bacteria obtain energy from the oxidation of ammonium or nitrite, while denitrifying bacteria use nitrate, nitrite, or nitrous oxide as terminal electron acceptors during the oxidation of organic substrates. Nitrate can also be used in a fermentative process by *Clostridium perfringens,* although since the process uses an inorganic electron acceptor, it is by definition a respiratory process.

The amounts of energy obtained by these processes are listed in Table. There seem to be very close couplings between nitrification and de-nitrification.

It has been postulated that the nitroxyl radical HNO is a possible intermediate in the oxidation of hydroxylamineto nitrite by *Nitrosomona.* There is still no direct evidence for this, however.

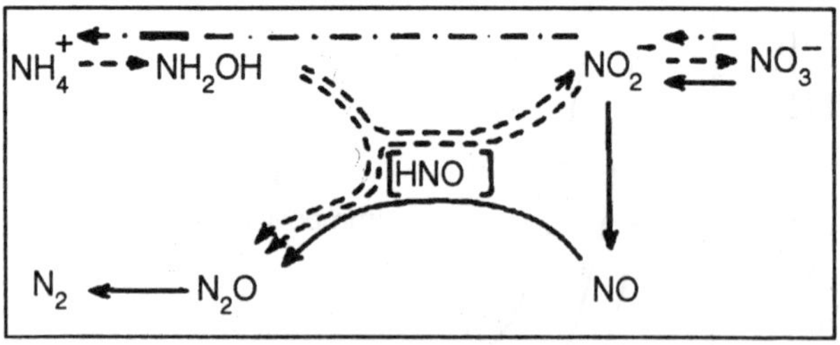

Fig. The Inorganic Nitrogen Cycle in Soil.

The presence of the nitroxyl radical would also explain the possibilities for N_2O production during nitrification, since, especially under anaerobic conditions, there would be a spontaneous (chemical) decomposition of [HNO] to N_2O. Nitroxyl also has been suggested by Kluyver and Verhoeven to be an intermediate in nitrite reduction by denitrifying bacteria, and since a dimerization step is entailed, this radical would seem to be a probable intermediate. *Nitrosomonas* also appears to reduce nitrite under anaerobic conditions.

From an ecological viewpoint, there may be several advantages:

- Nitrite may be used as a terminal electron acceptor during brief periods of anaerobiosis;

- Nitrite reduction may offer a detoxifying mechanism if nitrite concentrations become too high;
- Under certain circumstances nitrite reduction may minimize possible competition from *Nitrobacter* which is dependent on nitrite for growth.

Certain heterotrophic microorganisms may also oxidize ammonium to nitrate. Heterotrophic nitrification may be important in view of the possible ecological significance of some of the products identified.

Hydroxarnic acids may be important growth factors, possibly linked to iron uptake, and several of the identified metabolic products from heterotrophic nitrification are biocidal. Another interesting feature in the metabolism of inorganic nitrogen relates to the inhibitory effect of acetylene on several of the enzyme systems involved, viz. nitrogenase, ammonium oxidase, and nitrous oxide reductase. The effect on nitrification may complicate the use of the acetylene inhibition technique for determining denitrification in systems that are dependent on nitrate production by nitrifying bacteria.

NITROGEN FIXATION

Interest in nitrogen fixation has increased in recent years for several reasons. From a practical viewpoint, it was largely triggered by the tripling in the price of nitrogen fertilizers after 1974, and by the difficulties-especially for farmers in the less developed countries-of supplying nitrogen to the soil in sufficient amounts to increase crop production.

Much attention was accordingly given to optimizing the use of biological nitrogen fixation. The use of Rhizobium for inoculating legume crops constitutes the most promising approach and has been widely used for many years.

Significant research projects are being carried out in a large number of countries to optimize biological nitrogen fixation using Rhizobium. In tropical ecosystems the contribution of indigenous legumes is often more important that that of cultivated species, such as soy bean, peanut, cowpea and chick-pea. Specialized regional microbiological resource centres (MIRCENs) devoted to the use of biological nitrogen fixation have been set up in Kenya and Brazilunder the sponsorship of UNEP and Unesco. hnportant research efforts also are being made by the international agricultural research institutes, e.g.

International Institute for Tropical Agriculture (IITA) in Nigeria and the International Crop Research Institute for the Semi-arid Tropics (ICRISAT) in India Strain selection from native legumes will be important, especially the isolation of Rhizobium strains with an efficient hydrogenase, since strains which can recycle the hydrogen formed during nitrogen fixation have been shown to be more efficient in fixing nitrogen.

Attention has also been focused recently on the possibilities of associative symbioses between microorganisms and plant roots, the most important of which seems to be the association of Azospirillum with sorghum, maize, and sugar cane.

The association of Azotobacter paspali and Paspal-umnotatum is another such association that has received increased attention. The possibility that Azospirillum may denitrify and thus fix nitrogen under anaerobic conditions, with nitrate as a terminal electron acceptor, is another interesting recent flnding, which is a futher important link in the coupling of inorganic nitrogen metabolism in bacteria, as discussed above. The aquatic water fern Azolla, together with the blue-green alga Anabaena azollae, provides another example of a nitrogen-fixing association that can fix substantial amounts of nitrogen and be of agronomic significance. It is especially important in rice fields.

The recent isolation of the cyanobacterium responsible for the nitrogen fixation will be important in further attempts to investigate-and ultimately select-strains which are efficient in the symbiosis. An Azolia species which excretes ammonium during the entire growing season seems to be especially promising for agronomic use.

Non-legume symbiosis with actinomycetes of the genus Frankia occurs in several temperate species, such as Alnus glutinosa, Hippophaerhamnoides and Myricagale. Alnus species, for example, may become important in short-term, rotation forestry for energy production. Myrica gale is a common bog species in Europe and N. America and could be important in the nitrogen economy of such bog ecosystems.

Myrica can perhaps be used in the early stages of forest plantation after drainage. Recent reports of the successful isolation of Frankia from Alnus nodules and its growth in pure culture in laboratory media constitute an important breakthrough in the attempts to understand this symbiosis. Other nitrogen-fixing associations include leaf nodules of the angiosperm Gunnera, cycad-blue green algae root-nodule symbiosis, liverworts and lichens.New nitrogen-fixing species are, however, still being discovered, such as the Rhizobium symbiosis with the legume Aeschynomena indica, forming large numbers of stem nodules fixing nitrogen, and the herb Datiscacannabina, which has been shown to form root nodules of the Alnus type.

It is likely that a large number of nitrogen fixing species is as yet unidentified, some of which may prove useful in agriculture and forestry. For tropical ecosystems, it seems especially important to consider the use of legume trees in crop-rotation practices and multiple cropping systems, and their importance in savanna ecosystems has been stressed.

Interest has also been shown in the interaction between nitrogen-fixing bacteria and animals, one example being nitrogen fixation in the gut of termites. The bacterium responsible, Citrobacter freundii, has been isolated from termites, and it may be important in supplying combined nitrogen to the animals, which feed mainly on carbohydrate-rich material with a low nitrogen content. In the more distant future, the possibilities of introducing the genes (nif genes) for nitrogen fixation into new hosts that can be used for practical purposes is a fascinating prospect, even if there are still very large difficulties.

A MICRONICHE CONCEPT OF NITROGEN CYCLING

Most biological nitrogen transformations are brought

about by microorganisms, these processes being regulated both qualitatively and quantitatively by a large number of factors. The qualitative aspects are primarily regulated by oxygen and energy sources, while the quantitative aspects are regulated mainly by temperature and moisture.

For example, soil water has a dual function. Moisture is a rate-limiting factor for all biological processes. At moisture contents below 20 per cent, activity is generally low, while at higher moisture contents there is a rapid increase in activity; the process often reaches a maximal rate at about 80 per cent, after which it starts to decline.

Soil moisture regulates process rates and is thus an important quantitative factor. However, he often observed decreased rate at very high moisture contents reflects a qualitative change rather than a direct quantitative change.

At increased water contents, oxygen diffusion becomes limiting and the microbial metabolism switches from aerobic to anaerobic.

Water then acts as a regulator of the quality and only secondarily of the quantity, since anaerobic processes are generally slower than aerobic.

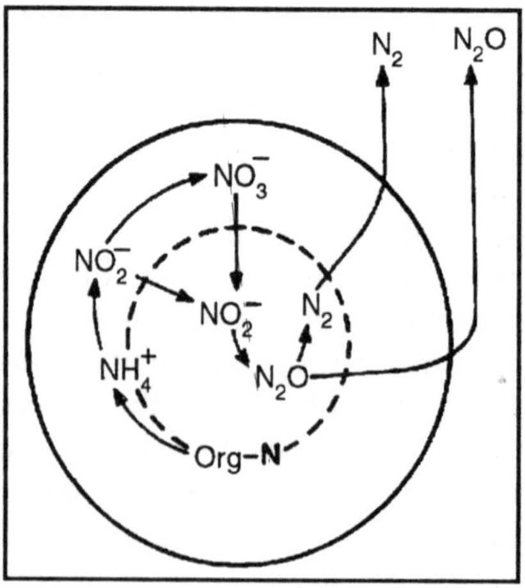

Fig. A Schematic Soil Crumb with an Aerobic outer Sector in which Nitrification can take Place and an Inner Anaerobic Centre from which Nitrogen may be Lost through De-Nitrification.

Water has a profound effect in the soil and makes generalizations of process rates (expressed as grams of nitrogen transformed per square metre) very difficult. As an example, let us consider the inorganic nitrogen metabolism of microorganisms.

A schematic representation of a soil crumb is shown in Figure, in which organic nitrogen is mineralized most rapidly in the aerobic zone; after nitrification, nitrate can diffuse into the anaerobic centre, where it can be denitrified, and gaseous nitrogen products are lost to the system. It has been estimated that most soil crumbs with a diameter of more than 3 mm have anaerobic centres.

It is also possible that *Nitrosomonas* bacteria can briefly switch to anaerobic metabolism, using nitrite as an electron acceptor, as previously discussed. Nitrogen-fixing aerobic bacteria may be able to use nitrate as a terminal electron acceptor and continue to fix nitrogen under anaerobic conditions, since it has been shown that *Rhizobium* and *Azospirillum* can denitrify.

It is possible that the nitrous oxide production observed from the nitrogen-fixing *Klebsiellapneumoniae* will be shown to be an end-product of de-nitrification, since it does not seem as if N_2O can be produced during assimilatory nitrite reduction. The occurrence of anaerobic microsites is probably also the reason for the observed De-nitrification rates in pelagic invertebrate fecal pellets.

The difficulties of determining the aerobic versus anaerobic activities in soils are linked to the difficulty of determining the extent of anaerobiosis. A schematic representation of a soil system with the various sensors which can be used for determining the aeration status of a soil, viz., an oxygen electrode, a redox electrode and a gas sampling probe.

Pt-micro electrodes with a diameter of less than one micron have been developed for use in live tissues, but a very large number of determinations must be made to determine the extent of anaerobiosis if such microsites are few. The difficulties of determining the different processes in the biogeochemical nitrogen cycle even on a square-metre basis are obvious. The difficulty of extrapolation, however, not only relates to space but also to time.

Single or even repeated measurements of, nitrogen fixation, for example, are very difficult to extrapolate to a yearly basis.

Soil bacteria show very rapid fluctuations in numbers just over a few days, especially after rainfall, a fact which provides another indication of the difficulty of extrapolating monthly rate determinations to yearly rates.

Frequent measurements must be made at the time of rainfall or irrigation as well as during high and low temperature regimesbefore realistic yearly rates can be calculated.

ECOSYSTEM NITROGEN CYCLES

The pattern of nitrogen cycling, together with that of the cycling of other nutrients, is an important factor in describing the functioning of ecosystems.

Odum postulated how nutrient cycling relates to the development of ecosystems. Detailed studies have been made on nitrogen cyclingin different ecosystems, but these have generally been made assuming steady-state conditions.

Only recently has attention been given to non-steady-state conditions such as comparisons of different successional stages, and impact of disturbances on ecosystem nitrogen cycles.

The concept of an increasing retention of available nitrogen during the successional development of a terrestrial ecosystem has been interpreted in a number of ways.

Haines suggested the following possible hypotheses:

- Roots in succeeding stages become more efficient in the uptake of dissolved and exchangeable nutrients;
- Increased soil organic matter in succeeding stages retains an increasing proportion of dissolved and exchangeable nutrients;
- A smaller proportion of the total nutrient pool of succeeding stages is lost from the system;

- In succeeding stages, a smaller proportion of the nutrient pool of the system is in flux at any given time.

It seems as if nitrate uptake decreases while ammonia uptake increases with succession. The possibility of an allelopathic effect of root exudates of climax species has been advanced as a possible explanation for the observed decrease in nitrate uptake, but there is only circumstantial evidence that this is the case.

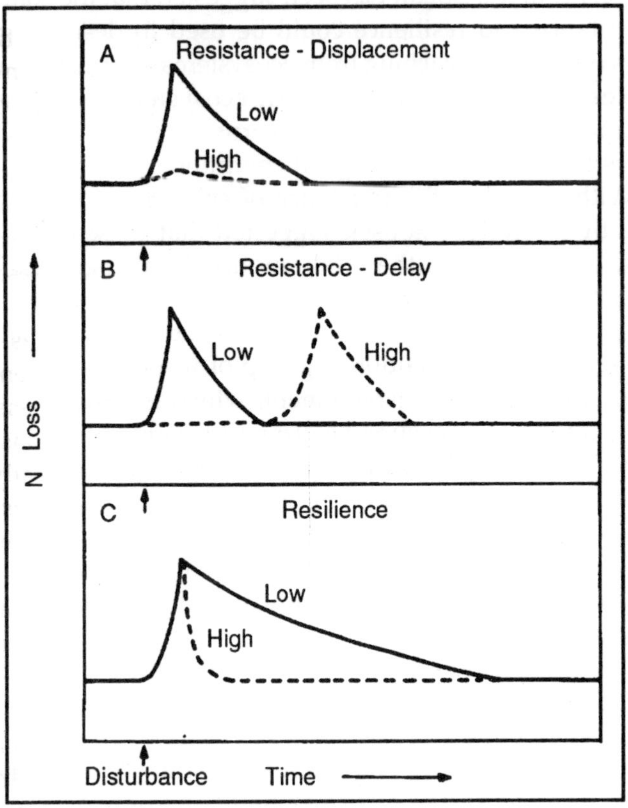

Fig. Relations Between Resistance and Resilience of Ecosystems to Disturbance and Subsequent Nitrogen Losses

Only by understanding what factors affect the various processes in the nitrogen cycle will it be possible to evaluate the merits of such an assumption. The fact that nitrate production can be accelerated in ecosystems subject to destructive disturbance was noted early on by Hesselrnan, and a number of studies on the magnitude of nitrogen losses following disturbance have been made; one of the major studies being carried out is that in the Hubbard Brook Experimental Forest.

The leaching of nitrogen from ecosystems following disturbance was recently investigated in a number of forest ecosystems in the USA.It has been suggested that the concepts of resistance and resilience could be used in describing the effects on nitrogen leaching from ecosystems after disturbance.

Resistance is the ability of an ecosystem to withstand perturbations, while resilience concerns the speed with which the system returns to its original condition.

With regard to the resistance of ecosystems to nitrogen losses by leaching, Vitousek suggested that these are of two types, one related to the total amount of nitrogen leached (displacement) and the other the resistance in time (delay). Using these terms, Vitousek proposed the following factors to be related to the mentioned group of responses:

- Resistance to displacement, which is related to rates of nitrogen cycling before perturbation;
- Delay, which is caused by nitrogen immobilization;
- Resilience, which is related to the rate of re-establishment of plant nitrogen uptake.

Most attention has been given the quantification of inputs and outputs of nitrogen in studies of the biogeochemical nitrogen cycle.

It should be noted, however, that the inputs and outputs of the nitrogen circulating in the vegetation- microorganism-soil system annually.

To understand what factors regulate transfers of nitrogen to and from ecosystems, further detailed knowledge of the processes involved in nitrogen transformations within the system is essential.

Data on ecosystem nitrogen budgets and an evaluation of the functional characteristics of the cycles are needed as important jigsaw pieces in the efforts to produce an authoritative and quantitative description of the global nitrogen cycle.

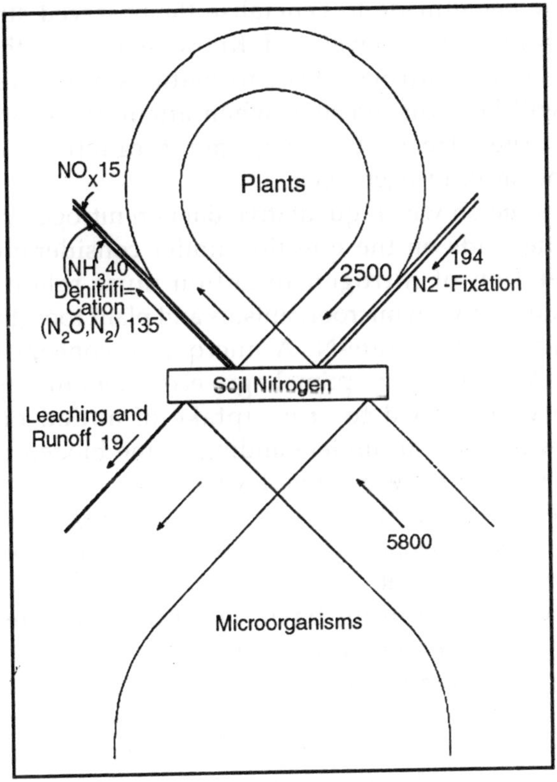

Fig. Annual Transfers of Nitrogen on a Global Basis Between Primary Producers and Microorganisms as Compared to Flows to and from the Terrestrial System. All flows in Tg yr^{-1} and Storages in Tg

Such an understanding is vital if we are to evaluate man's impact on the nitrogen cycle and the consequences that any disturbance might have.

THE GLOBAL NITROGEN CYCLE

Nitrification contribute to the global production of nitrous oxides, and do the oceans and terrestrial systems act both as a sink and source for nitrous oxide? We are slowly developing an ecological theory regarding nitrogen cycling in ecosystems and, based on an understanding of the individual processes, efforts are being made to generalize the observed strategies that ecosystems have developed. Knowledge of the theory of microbiological nitrogen transformations and ecosystem theory will be indispensable when attempts are made to forecast the effect of management practices on the biogeochemical nitrogen cycle.

The scale on which qualitative data on nitrogen cycles is needed depends on the question under consideration. An interpretation of nitrogen retention must relate to the competition between microorganisms and plant roots to utilize ammonium-N or nitrate-N. A microniche concept is then needed. Developing an optimal agro ecosystem management practice with regard to crop uptake of added nitrogen fertilizers calls for an understanding of the biogeochemical nitrogen cycle at the ecosystem level.

An evaluation of the possible effect of nitrogen oxides on the stratospheric ozone layer requires a quantification of the global nitrogen budget.

Only by combining basic laboratory studies with observations of nitrogen cycling dynamics in representative ecosystems-for example, as parts of the MAB programme- and by trying to integrate these data into a global quantitative model, will it be possible to meet the challenge for supplying the data badly needed for a cost-benefit analysis of fertilizer use and environmental risk assessments related to the increasing occurrence of nitrogen compounds as pollutants.

INTERACTIONS BETWEEN NITROGEN AND OTHER

Biogeochemical Cycles

The interactions between biogeochemical cycles are covered by other papers in this volume. It should be stressed that the biogeochemical nitrogen cycle should not be viewed in isolation, as there are close couplings with cycles of, for example, water, carbon, phosphorus, and sulphur. As mentioned above, the water cycle is instrumental in determining the fate of nitrogen in an ecosystem.

Water regulates the rates of the processes involved in Nitrogen transformations, it acts as a diffusion barrier for oxygen creating anaerobic conditions and it transports nitrogen to and from the ecosystem.

The bulk of soil nitrogen occurs in organic matter and mineralization of organic carbon leads to a release of ammonia nitrogen, although the relative proportions of carbon and nitrogen play an important regulatory role in net nitrogen mineralization as discussed above. When nitrogen fertilizers are added to a crop, photosynthesis and carbon immobilization ususally increase, but fertilization could also lead to increased soil organic matter decomposition with subsequent release of carbon dioxide.

Specific examples of the effect of phosphorus on nitrogen transformations are many. Nitrogen fixation and nitrification seem to be particularly sensitive to the availability of soil phosphorus. In P deficient soils, nitrogen fixation by legumes is increased considerably if the legume has an efficient VA-mycorrhizal symbiont, through which P uptake can improve.

In addition to the discussion above on the regulation of nitrification rates, Pancholy suggested that phosphorus deficiency was the main reason for restricted nitrification rates observed in phosphorus-poor savanna soils. Both nitrogen and sulphur compounds can acidify the environment and such changes in pH will affect nitrogen cycling processes. It is important to quantify the relative contributions of nitrogen and sulphur compounds to the possible acidification of soils and the fate of the acidifying compounds as they relate to the acid-base status. Nitrogen fixation and nitrification are particularly sensitive to reduced pH in soils.

Chapter 8

Carbon Cycle

CO_2 CONCENTRATION TRENDS

Before the Industrial Era, circa 1750, atmospheric carbon dioxide (CO_2) concentration was 280 ± 10 ppm for several thousand years. It has risen continuously since then, reaching 367 ppm in 1999.

The present atmospheric CO_2 concentration has not been exceeded during the past 420,000 years, and likely not during the past 20 million years. The rate of increase over the past century is unprecedented, at least during the past 20,000 years. The present atmospheric CO_2 increase is caused by anthropogenic emissions of CO_2.

About three-quarters of these emissions are due to fossil fuel burning. Fossil fuel burning (plus a small contribution from cement production) released on average 5.4 ± 0.3 PgC/yr during 1980 to 1989, and 6.3 ± 0.4 PgC/yr during 1990 to 1999. Land use change is responsible for the rest of the emissions.

The rate of increase of atmospheric CO_2 content was 3.3 ± 0.1 PgC/yr during 1980 to 1989 and 3.2 ± 0.1 PgC/yr during 1990 to 1999. These rates are less than the emissions, because some of the emitted CO2 dissolves in the oceans, and some is taken up by terrestrial ecosystems. Individual years show different rates of increase. For example, 1992 was low (1.9 PgC/yr), and 1998 was the highest (6.0 PgC/yr) since direct measurements began in 1957. This variability is mainly caused by variations in land and ocean uptake. Statistically, high rates of increase in atmospheric CO_2 have occurred in most El Niño years, although low rates occurred during the extended El

Niño of 1991 to 1994. Surface water CO_2 measurements from the equatorial Pacific show that the natural source of CO_2 from this region is reduced by between 0.2 and 1.0 PgC/yr during El Niño events, counter to the atmospheric increase.

It is likely that the high rates of CO_2 increase during most El Niño events are explained by reductions in land uptake, caused in part by the effects of high temperatures, drought and fire on terrestrial ecosystems in the tropics. Land and ocean uptake of CO_2 can now be separated using atmospheric measurements (CO_2, oxygen (O_2) and $^{13}CO_2$). For 1980 to 1989, the ocean-atmosphere flux is estimated as -1.9 ± 0.6 PgC/yr and the land-atmosphere flux as -0.2 ± 0.7 PgC/yr based on CO_2 and O_2 measurements (negative signs denote net uptake).

For 1990 to 1999, the ocean-atmosphere flux is estimated as -1.7 ± 0.5 PgC/yr and the land-atmosphere flux as -1.4 ± 0.7 PgC/yr. These figures are consistent with alternative budgets based on CO_2 and $^{13}CO_2$ measurements, and with independent estimates based on measurements of CO_2 and $^{13}CO_2$ in sea water.

The new 1980s estimates are also consistent with the ocean-model based carbon budget of the IPCC WGI Second Assessment Report. The new 1990s estimates update the budget derived using SAR methodologies for the IPCC Special Report on Land Use, Land Use Change and Forestry. The net CO_2 release due to land-use change during the 1980s has been estimated as 0.6 to 2.5 PgC/yr (central estimate 1.7 PgC/yr). This net CO^2 release is mainly due to deforestation in the tropics.

Uncertainties about land-use changes limit the accuracy of these estimates. Comparable data for the 1990s are not yet available. The land-atmosphere flux estimated from atmospheric observations comprises the *balance* of net CO_2 release due to land-use changes and CO_2 uptake by terrestrial systems (the "residual terrestrial sink"). The residual terrestrial sink is estimated as -1.9 PgC/yr (range -3.8 to +0.3 PgC/yr) during the 1980s.

It has several likely causes, including changes in land management practices and fertilisation effects of increased atmospheric CO_2 and nitrogen (N) deposition, leading to increased vegetation and soil carbon. Modelling based on atmospheric observations (inverse modelling) enables the land-atmosphere and ocean-atmosphere fluxes to be partitioned between broad latitudinal bands.

The sites of anthropogenic CO_2 uptake in the ocean are not resolved by inverse modelling because of the large, natural background airsea fluxes (outgassing in the tropics and uptake in high latitudes). Estimates of the land-atmosphere flux north of 30°N during 1980 to 1989 range from -2.3 to -0.6 PgC/yr; for the tropics, -1.0 to +1.5 PgC/yr.

These results imply substantial terrestrial sinks for anthropogenic CO_2 in the northern extra-tropics, and in the tropics (to balance deforestation). The pattern for the 1980s persisted into the 1990s. Terrestrial carbon inventory data indicate carbon sinks in northern and tropical forests, consistent with the results of inverse modelling.

East-west gradients of atmospheric CO_2 concentration are an order of magnitude smaller than north-south gradients. Estimates of continental-scale CO_2 balance are possible in principle but are poorly constrained because there are too few well-calibrated CO_2 monitoring sites, especially in the interior of continents, and insufficient data on air-sea fluxes and vertical transport in the atmosphere.

THE GLOBAL CARBON CYCLE AND ANTHROPOGENIC CO_2

The global carbon cycle operates through a variety of response and feedback mechanisms. The most relevant for decade to century time-scales are listed here.

Responses of the Carbon Cycle to Changing CO_2 Concentrations

- Uptake of anthropogenic CO_2 by the ocean is primarily governed by ocean circulation and carbonate chemistry. So long as atmospheric CO_2 concentration is increasing there is net uptake of carbon by the ocean, driven by the atmosphere-ocean difference in partial pressure of CO_2. The fraction of anthropogenic CO_2 that is taken up by the ocean declines with increasing CO_2 concentration, due to reduced buffer capacity of the carbonate system. The fraction taken up by the ocean also declines with the rate of increase of atmospheric CO_2, because the rate of mixing between deep water and surface water limits CO_2 uptake.

- Increasing atmospheric CO_2 has no significant fertilisation effect on marine biological productivity, but it decreases pH. Over a century, changes in marine biology brought about by changes in calcification at low pH could increase the ocean uptake of CO_2 by a few percentage points.
- Terrestrial uptake of CO_2 is governed by net biome production (NBP), which is the balance of net primary production (NPP) and carbon losses due to heterotrophic respiration (decomposition and herbivory) and fire, including the fate of harvested biomass. NPP increases when atmospheric CO_2 concentration is increased above present levels (the "fertilisation" effect occurs directly through enhanced photosynthesis, and indirectly through effects such as increased water use efficiency). At high CO_2 concentration (800 to 1,000 ppm) any further direct CO_2 fertilisation effect is likely to be small. The effectiveness of terrestrial uptake as a carbon sink depends on the transfer of carbon to forms with long residence times (wood or modified soil organic matter). Management practices can enhance the carbon sink because of the inertia of these "slow" carbon pools.

Feedbacks in the Carbon Cycle due to Climate Change

- Warming reduces the solubility of CO_2 and therefore reduces uptake of CO_2 by the ocean.
- Increased vertical stratification in the ocean is likely

to accompany increasing global temperature. The likely consequences include reduced outgassing of upwelled CO_2, reduced transport of excess carbon to the deep ocean, and changes in biological productivity.

- On short time-scales, warming increases the rate of heterotrophic respiration on land, but the extent to which this effect can alter land-atmosphere fluxes over longer timescales is not yet clear. Warming, and regional changes in precipitation patterns and cloudiness, are also likely to bring about changes in terrestrial ecosystem structure, geographic distribution and primary production. The net effect of climate on NBP depends on regional patterns of climate change.

Other Impacts on the Carbon Cycle

- Changes in management practices are very likely to have significant effects on the terrestrial carbon cycle. In addition to deforestation and afforestation/reforestation, more subtle management effects can be important. For example, fire suppression (e.g., in savannas) reduces CO_2 emissions from burning, and encourages woody plant biomass to increase. On agricultural lands, some of the soil carbon lost when land was cleared and tilled can be regained through adoption of low-tillage agriculture.
- Anthropogenic N deposition is increasing terrestrial NPP in some regions; excess tropospheric ozone (O_3) is likely to be reducing NPP.
- Anthropogenic inputs of nutrients to the oceans by rivers and atmospheric dust may influence marine biological productivity, although such effects are poorly quantified.

Modelling and Projection of CO_2 Concentration

Process-based models of oceanic and terrestrial carbon

cycling have been developed, compared and tested against *in situ* measurements and atmospheric measurements. The following are consistent results based on several models.

- Modelled ocean-atmosphere flux during 1980 to 1989 was in the range -1.5 to -2.2 PgC/yr for the 1980s, consistent with earlier model estimates and consistent with the atmospheric budget.
- Modelled land-atmosphere flux during 1980 to 1989 was in the range -0.3 to -1.5 PgC/yr, consistent with or slightly more negative than the land-atmosphere flux as indicated by the atmospheric budget. CO_2 fertilisation and anthropogenic N deposition effects contributed significantly: their combined effect was estimated as -1.5 to -3.1 PgC/yr. Effects of climate change during the 1980s were small, and of uncertain sign.
- In future projections with ocean models, driven by CO_2 concentrations derived from the IS92a scenario (for illustration and comparison with earlier work), ocean uptake becomes progressively larger towards the end of the century, but represents a smaller fraction of emissions than today. When climate change feedbacks are included, ocean uptake becomes less in all models, when compared with the situation without climate feedbacks.
- In analogous projections with terrestrial models, the rate of uptake by the land due to CO_2 fertilisation increases until mid-century, but the models project smaller increases, or no increase, after that time. When climate change feedbacks are included, land uptake becomes less in all models, when compared with the situation without climate feedbacks. Some models have shown a rapid decline in carbon uptake after the mid-century.

Two simplified, fast models (ISAM and Bern-CC) were used to project future CO_2 concentrations under IS92a and six SRES scenarios, and to project future emissions under five CO_2 stabilisation scenarios.

Both models represent ocean and terrestrial climate feedbacks, in a way consistent with processbased models, and allow for uncertainties in climate sensitivity and in ocean and terrestrial responses to CO_2 and climate.

- The reference case projections (which include climate feedbacks) of both models under IS92a are, by coincidence, close to those made in the SAR (which neglected feedbacks).
- The SRES scenarios lead to divergent CO_2 concentration trajectories. Among the six emissions scenarios considered, the projected range of CO_2 concentrations at the end of the century is 550 to 970 ppm (ISAM model) or 540 to 960 ppm (Bern-CC model).
- Variations in climate sensitivity and ocean and terrestrial model responses add at least 10 to +30% uncertainty to these values, and to the emissions implied by the stabilisation scenarios.
- The net effect of land and ocean climate feedbacks is always to increase projected atmospheric CO_2 concentrations. This is equivalent to reducing the allowable emissions for stabilisation at any one CO_2 concentration.
- New studies with general circulation models including interactive land and ocean carbon cycle components also indicate that climate feedbacks have the potential to increase atmospheric CO_2 but with large uncertainty about the magnitude of the terrestrial biosphere feedback.

IMPLICATIONS

CO_2 emissions from fossil fuel burning are virtually certain to be the dominant factor determining CO_2 concentrations during the 21st century. There is scope for land-use changes to increase or decrease CO_2 concentrations on this time-scale.

If all of the carbon so far released by land-use changes could be restored to the terrestrial biosphere, CO_2 at the end of the century would be 40 to 70 ppm less than it would be if no such intervention had occurred. By comparison, global deforestation would add two to four times more CO_2 to the atmosphere than reforestation of all cleared areas would subtract.

There is sufficient uptake capacity in the ocean to incorporate 70 to 80% of foreseeable anthropogenic CO_2 emissions to the atmosphere, this process takes centuries due to the rate of ocean mixing.

As a result, even several centuries after emissions occurred, about a quarter of the increase in concentration caused by these emissions is still present in the atmosphere.

CO_2 stabilisation at 450, 650 or 1,000 ppm would require global anthropogenic CO_2 emissions to drop below 1990 levels, within a few decades, about a century, or about two centuries respectively, and continue to steadily decrease thereafter. Stabilisation requires that net anthropogenic CO_2 emissions ultimately decline to the level of persistent natural land and ocean sinks, which are expected to be small (<0.2 PgC/yr).

CONCENTRATION OF CO_2

The concentration of CO_2 in the atmosphere has risen from close to 280 parts per million (ppm) in 1800, at first slowly and then progressively faster to a value of 367 ppm in 1999, echoing the increasing pace of global agricultural and industrial development. This is known from numerous, well-replicated measurements of the composition of air bubbles trapped in Antarctic ice.

Atmospheric CO_2 concentrations have been measured directly with high precision since 1957; these measurements agree with ice-core measurements, and show a continuation of the increasing trend up to the present.

Several additional lines of evidence confirm that the recent and continuing increase of atmospheric CO_2 content is caused by anthropogenic CO_2 emissions - most importantly fossil fuel burning.

First, atmospheric O_2 is declining at a rate comparable with fossil fuel emissions of CO_2 (combustion consumes O_2).

Second, the characteristic isotopic signatures of fossil fuel (its lack of ^{14}C, and depleted content of ^{13}C) leave their mark in the atmosphere.

Third, the increase in observed CO_2 concentration has been faster in the northern hemisphere, where most fossil fuel burning occurs.

Atmospheric CO_2 is, however, increasing only at about half the rate of fossil fuel emissions; the rest of the CO_2 emitted either dissolves in sea water and mixes into the deep ocean, or is taken up by terrestrial ecosystems.

Uptake by terrestrial ecosystems is due to an excess of primary production (photosynthesis) over respiration and other oxidative processes (decomposition or combustion of organic material).

Terrestrial systems are also an anthropogenic source of CO_2 when land-use changes (particularly deforestation) lead to loss of carbon from plants and soils.

Nonetheless, the global balance in terrestrial systems is currently a net uptake of CO_2. The part of fossil fuel CO_2 that is taken up by the ocean and the part that is taken up by the land can be calculated from the changes in atmospheric CO_2 and O_2 content because terrestrial processes of CO_2 exchange involve exchange of oxygen whereas dissolution in the ocean does not.

The human influence on the fluxes of carbon among the three"reservoirs" (atmosphere, ocean, and terrestrial biosphere) represent a small but significant perturbation of a huge global cycle.

This chapter summarises current knowledge of the global carbon cycle, with special reference to the fate of fossil fuel CO_2 and the factors that influence the uptake or release of CO_2 by the oceans and land.

These factors include atmospheric CO_2 concentration itself, the naturally variable climate, likely climate changes caused by increasing CO_2 and other greenhouse gases, changes in ocean circulation and biology, fertilising effects of atmospheric CO_2 and nitrogen deposition, and direct human actions such as land conversion (from native vegetation to agriculture and vice versa), fire suppression and land management for carbon storage as provided for by the Kyoto Protocol.

Any changes in the function of either the terrestrial biosphere or the ocean whether intended or not could potentially have significant effects, manifested within years to decades, on the fraction of fossil fuel CO_2 that stays in the atmosphere.

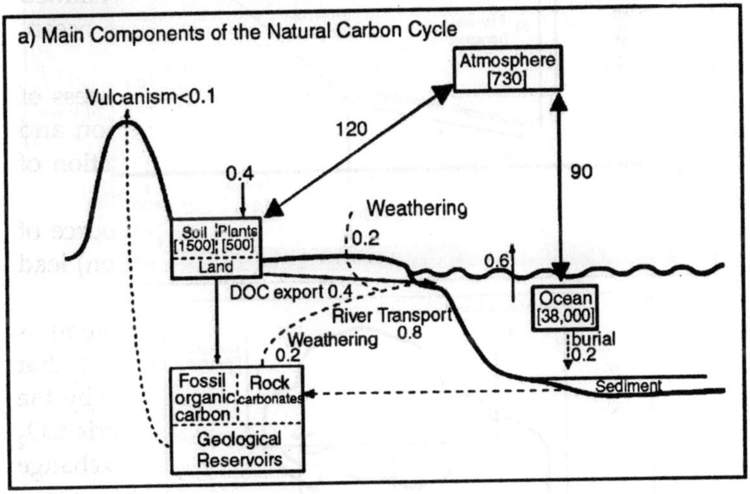

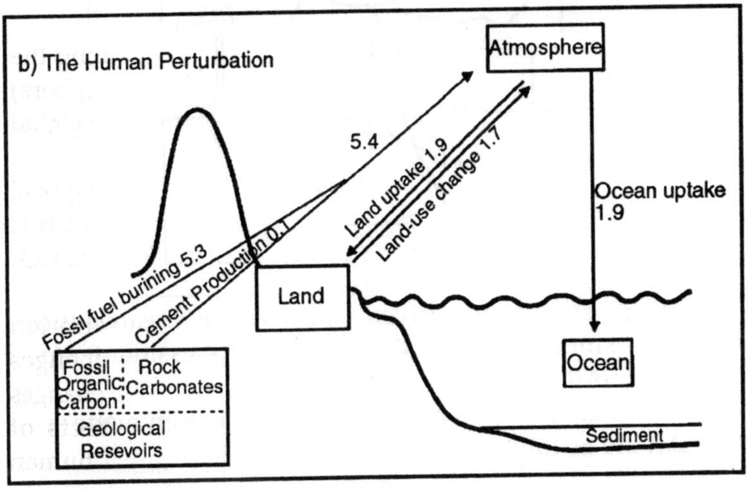

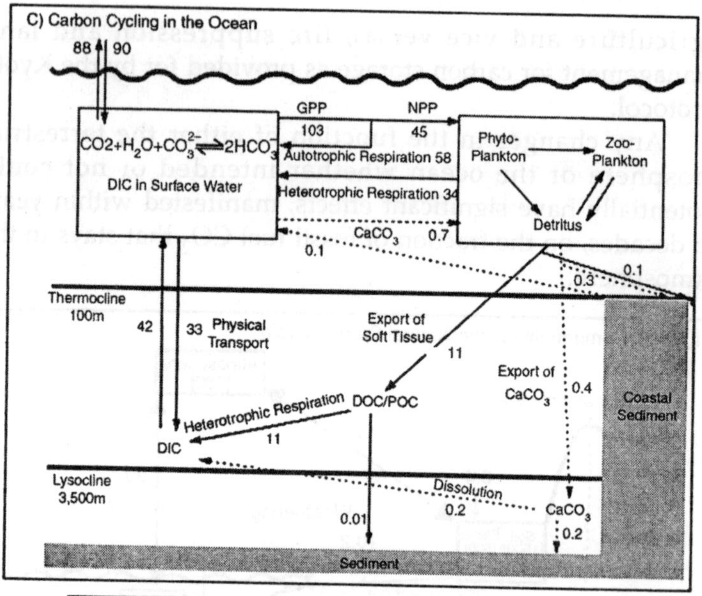

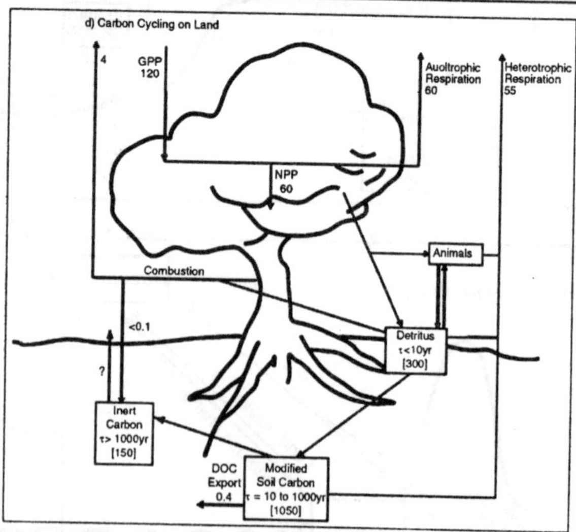

This perspective has driven a great deal of research during the years since the IPCC WGI Second Assessment report (hereafter SAR).

Some major areas where advances have been made since the SAR are as follows:

- Observational research (atmospheric, marine and terrestrial) aimed at a better quantification of carbon fluxes on local, regional and global scales. For example, improved precision and repeatability in atmospheric CO_2 and stable isotope measurements; the development of highly precise methods to measure changes in atmospheric O_2 concentrations; local terrestrial CO_2 flux measurements from towers, which are now being performed continuously in many terrestrial ecosystems; satellite observations of global land cover and change; and enhanced monitoring of geographical, seasonal and interannual variations of biogeochemical parameters in the sea, including measurements of the partial pressure of CO_2 (pCO_2) in surface waters.
- Experimental manipulations, for example: laboratory and greenhouse experiments with raised and lowered CO_2 concentrations; field experiments on ecosystems using free-air carbon dioxide enrichment (FACE) and open-top chamber studies of raised CO_2 effects, studies of soil warming and nutrient enrichment effects; and *in situ* fertilisation experiments on marine ecosystems and associated pCO_2 measurements.
- Theory and modelling, especially applications of atmospheric transport models to link atmospheric observations to surface fluxes (inverse modelling); the development of process-based models of terrestrial and marine carbon cycling and programmes to compare and test these models against observations; and the use of such models to project climate feedbacks on the uptake of CO_2 by the oceans and land.

As a result of this research, there is now a more firmly based knowledge of several central features of the carbon cycle.

For example:

- Time series of atmospheric CO_2, O_2 and $^{13}CO_2$ measurements have made it possible to observationally constrain the partitioning of CO_2

between terrestrial and oceanic uptake and to confirm earlier budgets, which were partly based on model results.

- *In situ* experiments have explored the nature and extent of CO_2 responses in a variety of terrestrial ecosystems (including forests), and have confirmed the existence of iron limitations on marine productivity.
- Process-based models of terrestrial and marine biogeochemical processes have been used to represent a complex array of feedbacks in the carbon cycle, allowing the net effects of these processes to be estimated for the recent past and for future scenarios.

MEASURING TERRESTRIAL CARBON STOCKS AND FLUXES

Estimating the carbon stocks in terrestrial ecosystems and accounting for changes in these stocks requires adequate information on land cover, carbon density in vegetation and soils, and the fate of carbon (burning, removals, decomposition). Accounting for changes in all carbon stocks in all areas would yield the net carbon exchange between terrestrial ecosystems and the atmosphere (NBP). Global land cover maps show poor agreement due to different definitions of cover types and inconsistent sources of data. Land cover changes are difficult to document, uncertainties are large, and historical data are sparse. Satellite imagery is a valuable tool for estimating land cover, despite problems with cloud cover, changes at fine spatial scales, and interpretation (for example, difficulties in distinguishing primary and secondary forest).

Aerial photography and ground measurements can be used to validate satellite-based observations. The carbon density of vegetation and soils has been measured in numerous ecological field studies that have been aggregated to a global scale to assess carbon stocks and NPP, although high spatial and temporal heterogeneity and methodological differences introduce large uncertainties. Land inventory studies tend to measure the carbon stocks in vegetation and soils over larger areas and/or longer time periods.

For example, the United Nations Food and Agricultural Organisation (FAO) has been compiling forest inventories since 1946 providing detailed data on carbon stocks, often based on commercial wood production data. Inventory studies include managed forests with mixed age stands, thus average carbon stock values are often lower than those based on ecological site studies, which have generally been carried out in relatively undisturbed, mature ecosystems.

Fluxes of carbon can be estimated from changes in inventoried carbon stocks, or from combining data on land-use change with methods to calculate changes in carbon stock. The greatest uncertainty in both methods is in estimating the fate of the carbon: the fraction which is burned, rates of decomposition, the effect of burning and harvesting on soil carbon, and subsequent land management. Ecosystem-atmosphere CO_2 exchange on short time-scales can be measured using micrometeorological techniques such as eddy covariance, which relies on rapidly responding sensors mounted on towers to resolve the net flux of CO_2 between a patch of land and the atmosphere.

The annual integral of the measured CO_2 exchange is approximately equivalent to NEP. This innovation has led to the establishment of a rapidly expanding network of long-term monitoring sites (FLUXNET) with many sites now operating for several years, improving the understanding of the physiological and ecological processes that control NEP.

The distribution of sites is currently biased toward regrowing forests in the Northern Hemisphere, and there are still technical problems and uncertainties, although these are being tackled. Current flux measurement techniques typically integrate processes at a scale less than 1 km^2.

TERRESTRIAL AND OCEAN BIOGEOCHEMISTRY

Overview of the Carbon Cycle

The first panel of Figure shows the major components of

the carbon cycle, estimates of the current storage in the active compartments, and estimates of the gross fluxes between compartments.

The second panel shows best estimates of the additional flux (release to the atmosphere – positive; uptake – negative) associated with the human perturbation of the carbon cycle during the 1980s. Note that the gross amounts of carbon annually exchanged between the ocean and atmosphere, and between the land and atmosphere, represent a sizeable fraction of the atmospheric CO_2 content – and are many times larger than the total anthropogenic CO_2 input.

In consequence, an imbalance in these exchanges could easily lead to an anomaly of comparable magnitude to the direct anthropogenic perturbation. This implies that it is important to consider how these fluxes may be changing in response to human activities.

To understand how the changing global environment may alter the carbon cycle, it is necessary to further analyse the fluxes and examine the physicochemical and biological processes that determine them.

The remaining two panels of Figure indicate the main constituent fluxes in the terrestrial and marine systems, with current estimates of their magnitude. The following sections explain the controls on these fluxes, with special reference to processes by which anthropogenic changes may influence the overall carbon balance of the land and oceans on time-scales from years to centuries.

Terrestrial Carbon Processes

Higher plants acquire CO_2 by diffusion through tiny pores (stomata) into leaves and thus to the sites of photosynthesis. The total amount of CO_2 that dissolves in leaf water amounts to about 270 PgC/yr, i.e., more than one-third of all the CO_2 in the atmosphere.

This quantity is measurable because this CO_2 has time to exchange oxygen atoms with the leaf water and is imprinted with the corresponding ^{18}O "signature". Most of this CO_2 diffuses out again without participating in photosynthesis. The amount that is "fixed" from the atmosphere, i.e., converted from CO_2 to carbohydrate during photosynthesis, is known as gross primary production (GPP).

Terrestrial GPP has been estimated as about 120 PgC/yr based on ^{18}O measurements of atmospheric CO_2. This is also the approximate value necessary to support observed plant growth, assuming that about half of GPP is incorporated into new plant tissues such as leaves, roots and wood, and the other half is converted back to atmospheric CO_2 by autotrophic respiration (respiration by plant tissues).

Annual plant growth is the difference between photosynthesis and autotrophic respiration, and is referred to as net primary production (NPP). NPP has been measured in all major ecosystem types by sequential harvesting or by measuring plant biomass. Global terrestrial NPP has been estimated at about 60 PgC/yr through integration of field measurements.

Estimates from remote sensing and atmospheric CO_2 data concur with this value, although there are large uncertainties in all methods. Eventually, virtually all of the carbon fixed in NPP is returned to the atmospheric CO_2 pool through two processes: heterotrophic respiration (Rh) by decomposers (bacteria and fungi feeding on dead tissue and exudates) and herbivores; and combustion in natural or human-set fires. Most dead biomass enters the detritus and soil organic matter pools where it is respired at a rate that depends on the chemical composition of the dead tissues and on environmental conditions (for example, low temperatures, dry conditions and flooding slow down decomposition).

Conceptually, several soil carbon pools are distinguished. Detritus and microbial biomass have a short turnover time (<10 yr). Modified soil organic carbon has decadal to centennial turnover time. Inert (stable or recalcitrant) soil organic carbon is composed of molecules more or less resistant to further decomposition.

A very small fraction of soil organic matter, and a small fraction of burnt biomass, are converted into inert forms. Natural processes and management regimes may reduce or increase the amount of carbon stored in pools with turnover times on the order of tens to hundreds of years (living wood, wood products and modified soil organic matter) and thus influence the time evolution of atmospheric CO_2 over the century. The difference between NPP and Rh determines how much carbon is lost or gained by the ecosystem in the absence of disturbances that remove carbon from the ecosystem (such as harvest or fire).

This carbon balance, or net ecosystem production (NEP), can be estimated from changes in carbon stocks, or by measuring the fluxes of CO_2 between patches of land and the atmosphere. Annual NEP flux measurements are in the range 0.7 to 5.9 MgC/ha/yr for tropical forests and 0.8 to 7.0 MgC/ha/yr for temperate forests; boreal forests can reach up to 2.5 MgC/ha/yr although they have been shown to be carbonneutral or to release carbon in warm and/or cloudy years. Integration of these and other results leads to an estimated global NEP of about 10 PgC/yr, although this is likely to be an overestimate because of the current biased distribution of flux measuring sites.

When other losses of carbon are accounted for, including fires, harvesting/removals (eventually combusted or decomposed), erosion and export of dissolved or suspended organic carbon (DOC) by rivers to the oceans, what remains is the net biome production (NBP), i.e., the carbon accumulated by the terrestrial biosphere. This is what the atmosphere ultimately "sees" as the net land uptake on a global scale over periods of a year or more. NBP is estimated in this chapter to have averaged -0.2 ± 0.7 PgC/yr during the 1980s and -1.4 ± 0.7 PgC/yr during the 1990s, based on atmospheric measurements of CO_2 and O_2.

By definition, for an ecosystem in steady state, Rh and other carbon losses would just balance NPP, and NBP would be zero. In reality, human activities, natural disturbances and climate variability alter NPP and Rh, causing transient changes in the terrestrial carbon pool and thus non-zero NBP.

If the rate of carbon input (NPP) changes, the rate of carbon output (Rh) also changes, in proportion to the altered carbon content; but there is a time lag between changes in NPP and changes in the slower responding carbon pools. For a *step* increase in NPP, NBP is expected to increase at first but to relax towards zero over a period of years to decades as the respiring pool "catches up".

The globally averaged lag required for Rh to catch up with a change in NPP has been estimated to be of the order of 10 to 30 years. A *continuous* increase in NPP is expected to produce a sustained positive NBP, so long as NPP is still increasing, so that the increased terrestrial carbon has not been processed through the respiring carbon pools, and provided that the increase is not outweighed by compensating increases in mortality or disturbance.

The terrestrial system is currently acting as a global sink for carbon despite large releases of carbon due to deforestation in some regions. Likely mechanisms for the sink are known, but their relative contribution is uncertain.

Natural climate variability and disturbance regimes (including fire and herbivory) affect NBP through their impacts on NPP, allocation to long- versus short-lived tissues, chemical and physical properties of litter, stocks of living biomass, stocks of detritus and soil carbon, environmental controls on decomposition and rates of biomass removal.

Human impacts occur through changes in land use and land management, and through indirect mechanisms including climate change, and fertilisation due to elevated CO_2 and deposition of nutrients (most importantly, reactive nitrogen). These mechanisms are discussed individually in the following sections.

Effects of Changes in Land use and Land Management

Changes in land use and management affect the amount of carbon in plant biomass and soils. Historical cumulative carbon losses due to changes in land use have been estimated to be 180 to 200 PgC by comparing maps of "natural" vegetation in the absence of human disturbance (derived from ground-based information or from modelled potential vegetation based on climate) to a map of current vegetation derived from 1987 satellite data.

Houghton estimated emissions of 121 PgC (approximately 60% in tropical areas and 40% in temperate areas) for the period 1850 to 1990 from statistics on land-use change, and a simple model tracking rates of decomposition from different pools and rates of regrowth on abandoned or reforested land.

There was substantial deforestation in temperate areas prior to 1850, and this may be partially reflected in the difference between these two analyses. The estimated land-use emissions during 1850 to 1990 of 121 PgC can be compared to estimated net terrestrial flux of 39 PgC to the atmosphere over the same period inferred from an atmospheric increase of 144 PgC, a release of 212 PgC due to fossil fuel burning, and a modelled ocean-atmosphere flux of about 107 PgC.

The difference between the net terrestrial flux and estimated land-use change emissions implies a residual land-atmosphere flux of 82 PgC (i.e., a terrestrial sink) over the same period. Land use responds to social and economic pressures to provide food, fuel and wood products, for subsistence use or for export.

Land clearing can lead to soil degradation, erosion and leaching of nutrients, and may therefore reduce the subsequent ability of the ecosystem to act as a carbon sink. Ecosystem conservation and management practices can restore, maintain and enlarge carbon stocks.

Fire is important in the carbon budget of some ecosystems (e.g., boreal forests, grasslands, tropical savannas and woodlands) and is affected directly by management and indirectly by land-use change. Fire is a major short-term source of carbon, but adds to a small longer-term sink (<0.1 PgC/yr) through production of slowly decomposing and inert black carbon.

FORESTS

Deforestation has been responsible for almost 90% of the estimated emissions due to land-use change since 1850, with a 20% decrease of the global forest area. Deforestation appears to be slowing slightly in tropical countries, and some deforested areas in Europe and North America have been reforested in recent decades. Managed or regenerated forests generally store less carbon than natural forests, even at maturity. New trees take up carbon rapidly, but this slows down towards maturity when forests can be slight sources or sinks. To use land continuously in order to take up carbon, the wood must be harvested and turned into long-lived products and trees must be re-planted. The trees may also be used for biomass energy to avoid future fossil fuel emissions.

Analysis of scenarios for future development show that expanded use of biomass energy could reduce the rate of atmospheric CO_2 increase estimated that slowing deforestation and promoting natural forest regeneration and afforestation could increase carbon stocks by about 60 to 87 PgC over the period 1995 to 2050, mostly in the tropics.

Savannas and Grasslands – fire and Grazing

Grasslands and mixed tree-grass systems are vulnerable to subtle environmental and management changes that can lead to shifts in vegetation state. Livestock grazing on these lands is the land use with the largest global areal extent. Extensive clearing of trees (for agricultural expansion) has occurred in some areas.

In other areas, fire suppression, eradication of indigenous browsers and the introduction of intensive grazing and exotic trees and shrubs have caused an increase in woody plant density known as woody encroachment or tree thickening. This process has been estimated to result in a CO_2 sink of up to 0.17 PgC/yr in the USA during the 1980s and at least 0.03 PgC/yr in Australia. Grassland ecosystems have high root production and store most of their carbon in soils where turnover is relatively slow, allowing the possibility of enhancement through management.

Peatlands/wetlands

Peatlands/wetlands are large reserves of carbon, because

anaerobic soil conditions and (in northern peatlands) low temperatures reduce decomposition and promote accumulation of organic matter.

Total carbon stored in northern peatlands has been estimated as about 455 PgC with a current uptake rate in extant northern peatlands of 0.07 PgC/yr. Anaerobic decomposition releases methane (CH_4) which has a global warming potential (GWP) about 23 times that of CO_2.

The balance between CH_4 release and CO_2 uptake and release is highly variable and poorly understood. Draining peatlands for agriculture increases total carbon released by decomposition, although less is in the form of CH_4. Forests grown on drained peatlands may be sources or sinks of CO_2 depending on the balance of decomposition and tree growth.

Agricultural Land

Conversion of natural vegetation to agriculture is a major source of CO_2, not only due to losses of plant biomass but also, increased decomposition of soil organic matter caused by disturbance and energy costs of various agricultural practices. Conversely, the use of highyielding plant varieties, fertilisers, irrigation, residue management and reduced tillage can reduce losses and enhance uptake within managed areas.

These processes have led to an estimated increase of soil carbon in agricultural soils in the USA of 0.14 PgC/yr during the 1980s. IPCC estimated that appropriate management practices could increase carbon sinks by 0.4 to 0.9 PgC/yr , or a cumulative carbon storage of 24 to 43 PgC over 50 years; energy efficiency improvements and production of energy from dedicated crops and residues would result in a further mitigation potential of 0.3 to 1.4 PgC/yr, or a cumulative carbon storage of 16 to 68 PgC over 50 years.

Scenarios

The IPCC Special Report on Land Use, Land-Use Change and Forestry (hereafter SRLULUCF) derived scenarios of land-use emissions for the period 2008 to 2012. It was estimated that a deforestation flux of 1.79 PgC/yr is likely to be offset by reforestation and afforestation flux of -0.20 to -0.58 PgC/yr, yielding a net release of 1.59 to 1.20 PgC/yr.

The potential for net carbon storage from several "additional activities" such as improved land management and other land-use changes was estimated to amount to a global land-atmosphere flux in the region of -1.3 PgC/yr in 2010 and -2.5 PgC/yr in 2040, not including wood products and bioenergy.

EFFECTS OF CLIMATE

Solar radiation, temperature and available water affect photosynthesis, plant respiration and decomposition, thus climate change can lead to changes in NEP. A substantial part of the interannual variability in the rate of increase of CO_2 is likely to reflect terrestrial biosphere responses to climate variability.

Warming may increase NPP in temperate and arctic ecosystems where it can increase the length of the seasonal and daily growing cycles, but it may decrease NPP in waterstressed ecosystems as it increases water loss. Respiratory processes are sensitive to temperature; soil and root respiration have generally been shown to increase with warming in the short term although evidence on longer-term impacts is conflicting.

Changes in rainfall pattern affect plant water availability and the length of the growing season, particularly in arid and semi-arid regions. Cloud cover can be beneficial to NPP in dry areas with high solar radiation, but detrimental in areas with low solar radiation. Changing climate can also affect the distribution of plants and the incidence of disturbances such as fire (which could increase or decrease depending on warming and precipitation patterns, possibly resulting under some circumstances in rapid losses of carbon), wind, and insect and pathogen attacks, leading to changes in NBP.

The global balance of these positive and negative effects of climate on NBP depends strongly on regional aspects of climate change. The climatic sensitivity of high northern latitude ecosystems (tundra and taiga) has received particular attention as a consequence of their expanse, high carbon density, and observations of disproportionate warming in these regions.

High-latitude ecosystems contain about 25% of the total world soil carbon pool in the permafrost and the seasonally-thawed soil layer. This carbon storage may be affected by changes in temperature and water table depth. High latitude ecosystems have low NPP, in part due to short growing seasons, and slow nutrient cycling because of low rates of decomposition in waterlogged and cold soils.

Remotely sensed data and phenological observations independently indicate a recent trend to longer growing seasons in the boreal zone and temperate Europe. Such a trend might be expected to have increased annual NPP.

A shift towards earlier and stronger spring depletion of atmospheric CO_2 has also been observed at northern stations, consistent with earlier onset of growth at mid- to high northern latitudes. However, recent flux measurements at individual high-latitude sites have generally failed to find appreciable NEP. These studies suggest that, at least in the short term, any direct effect of warming on NPP may be more than offset by an increased respiration of soil carbon caused by the effects of increased depth of soil thaw. Increased decomposition, may, however also increase nutrient mineralisation and thereby indirectly stimulate NPP.

Large areas of the tropics are arid and semi-arid, and plant production is limited by water availability. There is evidence that even evergreen tropical moist forests show reduced GPP during the dry season and may become a carbon source under the hot, dry conditions of typical El Niño years. With a warmer ocean surface, and consequently generally increased precipitation, the global trend in the tropics might be expected to be towards increased NPP, but changing precipitation patterns could lead to drought, reducing NPP and increasing fire frequency in the affected regions.

Effects of Increasing Atmospheric CO_2

CO_2 and O_2 compete for the reaction sites on the

photosynthetic carbon-fixing enzyme, Rubisco. Increasing the concentration of CO_2 in the atmosphere has two effects on the Rubisco reactions: increasing the rate of reaction with CO_2 (carboxylation) and decreasing the rate of oxygenation. Both effects increase the rate of photosynthesis, since oxygenation is followed by photorespiration which releases CO_2.

With increased photsynthesis, plants can develop faster, attaining the same final size in less time, or can increase their final mass. In the first case, the overall rate of litter production increases and so the soil carbon stock increases; in the second case, both the below-ground and above-ground carbon stocks increase.

Both types of growth response to elevated CO_2 have been observed. The strength of the response of photosynthesis to an increase in CO_2 concentration depends on the photosynthetic pathway used by the plant.

Plants with a photosynthetic pathway known as C_3 (all trees, nearly all plants of cold climates, and most agricultural crops including wheat and rice) generally show an increased rate of photosynthesis in response to increases in CO_2 concentration above the present level. Plants with the C_4 photosynthetic pathway (tropical and many temperate grasses, some desert shrubs, and some crops including maize and sugar cane) already have a mechanism to concentrate CO_2 and therefore show either no direct photosynthetic response, or less response than C_3 plants.

Increased CO_2 has also been reported to reduce plant respiration under some conditions, although this effect has been questioned. Increased CO_2 concentration allows the partial closure of stomata, restricting water loss during transpiration and producing an increase in the ratio of carbon gain to water loss ("water-use efficiency", WUE).

This effect can lengthen the duration of the growing season in seasonally dry ecosystems and can increase NPP in both C_3 and C_4 plants. Nitrogen-use efficiency also generally improves as carbon input increases, because plants can vary the ratio between carbon and nitrogen in tissues and require lower concentrations of photosynthetic enzymes in order to carry out photosynthesis at a given rate; for this reason, low nitrogen availability does not consistently limit plant responses to increased atmospheric CO_2.

Increased CO_2 concentration may also stimulate nitrogen fixation. Changes in tissue nutrient concentration may affect herbivory and decomposition, although long-term decomposition studies have shown that the effect of elevated CO_2 in this respect is likely to be small because changes in the C:N ratio of leaves are not consistently reflected in the C:N ratio of leaf litter due to nitrogen retranslocation. The process of CO_2 "fertilisation" thus involves direct effects on carbon assimilation and indirect effects such as those via water saving and interactions between the carbon and nitrogen cycles.

Increasing CO_2 can therefore lead to structural and physiological changes in plants and can further affect plant competition and distribution patterns due to responses of different species. Field studies show that the relative stimulation of NPP tends to be greater in low-productivity years, suggesting that improvements in water- and nutrient-use efficiency can be more important than direct NPP stimulation.

Although NPP stimulation is not automatically reflected in increased plant biomass, additional carbon is expected to enter the soil, via accelerated ontogeny, which reduces lifespan and results in more rapid shoot death, or by enhanced root turnover or exudation. Because the soil microbial community is generally limited by the availability of organic substrates, enhanced addition of labile carbon to the soil tends to increase heterotrophic respiration unless inhibited by other factors such as low temperature.

Field studies have indicated increases in soil organic matter, and increases in soil respiration of about 30%, under elevated CO_2. The potential role of the soil as a carbon sink under elevated CO_2 is crucial to understanding NEP and long-term carbon dynamics, but remains insufficiently well understood.

C_3 crops show an average increase in NPP of around 33% for a doubling of atmospheric CO_2. Grassland and crop studies combined show an average biomass increase of 14%, with a wide range of responses among individual studies. In cold climates, low temperatures restrict the photosynthetic response to elevated CO_2. In tropical grasslands and savannas, C_4 grasses are dominant, so it has been assumed that trees and C_3 grasses would gain a competitive advantage at high CO_2. This is supported by carbon isotope evidence from the last glacial maximum, which suggests that low CO_2 favours C_4 plants.

However, field experiments suggest a more complex picture with C_4 plants sometimes doing better than C_3 under elevated CO_2 due to improved WUE at the ecosystem level. Highly productive forest ecosystems have the greatest potential for *absolute* increases in productivity due to CO_2 effects. Long-term field studies on young trees have typically shown a stimulation of photosynthesis of about 60% for a doubling of CO_2.

A FACE experiment in a fast growing young pine forest showed an increase of 25% in NPP for an increase in atmospheric CO_2 to 560 ppm. Some of this additional NPP is allocated to root metabolism and associated microbes; soil CO_2 efflux increases, returning a part (but not all) of the extra NPP to the atmosphere. The response of mature forests to increases in atmospheric CO_2 concentration has not been shown experimentally; it may be different from that of young forests for various reasons, including changes in leaf C:N ratios and stomatal responses to water vapour deficits as trees mature.

At high CO_2 concentrations there can be no further increase in photosynthesis with increasing CO_2, except through further stomatal closure, which may produce continued increases in WUE in water-limited environments. The shape of the response curve of global NPP at higher CO_2 concentrations than present is uncertain because the response at the level of gas exchange is modified by incompletely understood plantand ecosystem-level processes.

Based on photosynthetic physiology, it is likely that the additional carbon that could be taken up globally by enhanced photosynthsis as a direct consequence of rising atmospheric

CO_2 concentration is small at atmospheric concentrations above 800 to 1,000 ppm. Experimental studies indicate that some ecosystems show greatly reduced CO_2 fertilisation at lower concentrations than this.

Effects of Anthropogenic Nitrogen Deposition

Nitrogen availability is an important constraint on NPP, although phosphorus and calcium may be more important limiting nutrients in many tropical and subtropical regions. Reactive nitrogen is released into the atmosphere in the form of nitrogen oxides (NOx) during fossil fuel and biomass combustion and ammonia emitted by industrial regions, animal husbandry and fertiliser use. This nitrogen is then deposited fairly near to the source, and can act as a fertiliser for terrestrial plants.

There has been a rapid increase in reactive nitrogen deposition over the last 150 years. Much field evidence on nitrogen fertilisation effects on plants supports the hypothesis that additional nitrogen deposition will result in increased NPP, including the growth of trees in Europe. There is also evidence that N fertilisation enhances the formation of modified soil organic matter and thus increases the residence time of carbon in soils.

Tracer experiments with addition of the stable isotope ^{15}N provide insight into the short-term fate of deposited reactive nitrogen. It is clear from these experiments that most of the added N added to the soil surface is retained in the ecosystem rather than being leached out via water transport or returned to the atmosphere in gaseous form (as N_2, NO, N_2O or NH_3).

Studies have also shown that the tracer is found initially in the soil, but that it enters the vegetation after a few years. There is an upper limit to the amount of added N that can fertilise plant growth. This limit is thought to have been reached in highly polluted regions of Europe. With nitrogen saturation, ecosystems are no longer able to process the incoming nitrogen deposition, and may also suffer from deleterious effects of associated pollutants such as ozone (O_3), nutrient imbalance, and aluminium toxicity.

Additional Impacts of Changing Atmospheric Chemistry

Current tropospheric O_3 concentrations in Europe and North America cause visible leaf injury on a range of crop and tree species and have been shown to reduce the growth and yield of crops and young trees in experimental studies. The longer-term effects of O_3 on forest productivity are less certain, although significant negative associations between ozone exposure and forest growth have been reported in North America and in central Europe. O_3 is taken up through stomata, so decreased stomatal conductance at elevated CO_2 may reduce the effects of O_3.

There is also evidence of significant interactions between O_3 and soil water availability in effects on stem growth or NPP from field studies and from modelling studies. The regional impacts of O_3 on NPP elsewhere in the world are uncertain, although significant impacts on forests have been reported close to major cities. Fowler *et al.* estimate that the proportion of global forests exposed to potentially damaging ozone concentrations will increase from about 25% in 1990 to about 50% in 2100.

Other possible negative effects of industrially generated pollution on plant growth include effects of soil acidification due to deposition of,

$$NO_3^- \text{ and } SO_4^{2-}$$

severe forest decline has been observed in regions with high sulphate deposition, for instance in parts of eastern Europe and southern China. The wider effects are less certain and depend on soil sensitivity.

Fowler *et al.* estimate that 8% of global forest cover received an annual sulphate deposition above an estimated threshold for effects on acid sensitive soils, and that this will increase to 17% in 2050. The most significant long-term effect of continued acid deposition for forest productivity may be through depletion of base cations, with evidence of both increased leaching rates and decreased foliar concentrations, although the link between these changes in nutrient cycles and NPP needs to be quantified.

Additional Constraints on Terrestrial CO_2 Uptake

It is very likely that there are upper limits to carbon storage in ecosystems due to mechanical and resource constraints on the amount of above ground biomass and physical limits to the amount of organic carbon that can be held in soils. It is also generally expected that increased above-ground NPP (production of leaves and stem) will to some extent be counterbalanced by an increased rate of turnover of the biomass as upper limits are approached.

OCEAN CARBON PROCESSES

The total amount of carbon in the ocean is about 50 times greater than the amount in the atmosphere, and is exchanged with the atmosphere on a time-scale of several hundred years. Dissolution in the oceans provides a large sink for anthropogenic CO_2, due in part to its high solubility, but above all because of its dissociation into ions and interactions with sea water constituents. The annual two-way gross exchange of CO_2 between the atmosphere and surface ocean is about 90 PgC/yr, mediated by molecular diffusion across the air-sea interface.

Net CO_2 transfer can occur whenever there is a partial pressure difference of CO_2 across this interface. The flux can be estimated as the product of a gas transfer coefficient, the solubility of CO_2, and the partial pressure difference of CO_2 between air and water.

The gas transfer coefficient incorporates effects of many physical factors but is usually expressed as a non-linear function of wind speed alone. There is considerable uncertainty about this function.

Improvements in the ability to measure CO2 transfer directly may lead to a better knowledge of gas transfer coefficients. Despite extensive global measurements conducted during the 1990s, measurements of surface water pCO_2 remain sparse, and extensive spatial and temporal interpolation is required in order to produce global fields. Takahashi *et al.* interpolated data collected over three decades in order to derive monthly values of surface water pCO_2 over the globe for a single "virtual" calendar year. A wind speed dependent gas transfer coefficient was used to calculate monthly net CO_2 fluxes.

The resulting estimates, although subject to large uncertainty, revealed clear regional and seasonal patterns in net fluxes. Regional net CO_2 transfers estimated from contemporary surface water pCO_2 data should not be confused with the uptake of anthropogenic CO_2.

The uptake of anthropogenic CO_2 is the *increase* in net transfer over the pre-industrial net transfer, and is therefore superimposed on a globally varying pattern of relatively large natural transfers. The natural transfers result from heating and cooling, and biological production and respiration.

Carbon is transferred within the ocean from natural sink regions to natural source regions via ocean circulation and the sinking of carbon rich particles. This spatial separation of natural sources and sinks dominates the regional distribution of net annual air-sea fluxes.

CO_2 solubility is temperature dependent, hence air-sea heat transfer contributes to seasonal and regional patterns of air-sea CO_2 transfer. Net cooling of surface waters tends to drive CO_2 uptake; net warming drives outgassing. Regions of cooling and heating are linked via circulation, producing vertical gradients and north-south transports of carbon within the ocean (e.g., of the order 0.5 to 1 PgC/yr southward transport in the Atlantic Basin).

Biological processes also drive seasonal and regional distributions of CO_2 fluxes. The gross primary production by ocean phytoplankton has been estimated by Bender *et al.* to be 103 PgC/yr. Part of this is returned to DIC through autotrophic respiration, with the remainder being net primary production, estimated on the basis of global remote sensing data to be about 45 PgC/yr.

About 14 to 30% of the total NPP occurs in coastal areas. The resulting organic carbon is consumed by zooplankton (a quantitatively more important process than herbivory on land) or becomes detritus. Some organic carbon is released in dissolved form (DOC) and oxidised by bacteria with a fraction entering the ocean reservoir as net DOC production. Sinking of particulate organic carbon (POC) composed of dead organisms and detritus together with vertical transfer of DOC create a downward flux of organic carbon from the upper ocean known as "export production". Recent estimates for global export production range from roughly 10 to 20 PgC/yr.

An alternative estimate for global export production of 11 PgC/yr has been derived using an inverse model of physical and chemical data from the world's oceans.

Only a small fraction (about 0.1 PgC) of the export production sinks in sediments, mostly in the coastal ocean. Heterotrophic respiration at depth converts the remaining organic carbon back to DIC.

Eventually, and usually at another location, this DIC is upwelled into the ocean's surface layer again and may re-equilibrate with the atmospheric CO_2.

These mechanisms, often referred to as the biological pump, maintain higher DIC concentrations at depth and cause atmospheric CO_2 concentrations to be about 200 ppm lower than would be the case in the absence of such mechanisms. Marine organisms also form shells of solid calcium carbonate ($CaCO_3$) that sink vertically or accumulate in sediments, coral reefs and sands. This process depletes surface CO_3, reduces alkalinity, and tends to increase pCO_2 and drive more outgassing of CO_2.

The effect of $CaCO_3$ formation on surface water pCO_2 and air-sea fluxes is therefore counter to the effect of organic carbon production. For the surface ocean globally, the ratio between the export of organic carbon and the export of calcium carbonate (the "rain ratio") is a critical factor controlling the overall effect of biological activity on surface ocean pCO_2.

Milliman estimated a global production of $CaCO_3$ of 0.7 PgC/yr, with roughly equivalent amounts produced in shallow water and surface waters of the deep ocean. Of this total, approximately 60% accumulates in sediments. The rest re-dissolves either in the water column or within the sediment. An estimate of $CaCO_3$ flux analogous to the export production of organic carbon, however, should include sinking out of the upper layers of the open ocean, net accumulation in shallow sediments and reefs, and export of material from shallow systems into deep sea environments. Based on Milliman's budget, this quantity is about 0.6 PgC/yr (± 25 to 50 % at least).

The global average rain ratio has been variously estimated from models of varying complexity to be, 3.5 to 7.5, and 11. (It should be noted that rain ratios are highly depth dependent due to rapid oxidation of organic carbon at shallow depth compared to the depths at which sinking $CaCO_3$ starts to dissolve.) If one accepts an organic carbon export production value of 11 PgC/yr , then only Yamanaka and Tajika's value for the rain ratio approaches consistency with the observationbased estimates of the export of $CaCO_3$ and organic carbon from the ocean surface layer.

The overall productivity of the ocean is determined largely by nutrient supply from deep water. There are multiple potentially limiting nutrients: in practice nitrate and/or phosphate are commonly limiting. Silicate plays a role in limiting specific types of phytoplankton and hence in determining the qualitative nature of primary production, and potentially the depth to which organic carbon sinks.

A role for iron in limiting primary productivity in regions with detectable phosphate and nitrate but low productivity (HNLC or "high nutrient, low chlorophyll regions") has been experimentally demonstrated in the equatorial Pacific and the Southern Ocean. In both regions artificial addition of iron stimulated phytoplankton growth, resulting in decreased surface-water pCO_2.

In HLNC regions, the supply of iron from deep water, while an important source, is generally insufficient to meet the requirements of phytoplankton. An important additional supply of iron to surface waters far removed from sediment and riverine sources is aeolian transport and deposition. This aeolian supply of iron may limit primary production in HNLC regions, although the effect is ultimately constrained by the availability of nitrate and phosphate. Iron has been hypothesised to play an indirect role over longer time-scales (e.g., glacial-interglacial) through limitation of oceanic nitrogen fixation and, consequently, the oceanic content of nitrate.

The regional variability of oceanic nitrogen fixation and its temporal variability and potential climate-sensitivity have recently become apparent based on results from long time-series and global surveys. Carbon (organic and inorganic) derived from land also enters the ocean via rivers as well as to some extent via groundwater. This transport comprises a natural carbon transport together with a significant anthropogenic perturbation.

The global natural transport from rivers to the ocean is about 0.8 PgC/yr, half of which is organic and half inorganic. Additional fluxes due to human activity have been estimated to be about 0.1 PgC/yr (mainly organic carbon). Much of the organic carbon is deposited and/or respired and outgassed close to land, mostly within estuaries.

The outgassing of anthropogenic carbon from estuaries can be a significant term in comparison with regional CO_2 emissions estimates (e.g., 5 to 10% for Western Europe). The natural DIC transport via rivers, however, is part of a large-scale cycling of carbon between the open ocean and land associated with dissolution and precipitation of carbonate minerals.

This natural cycle drives net outgassing from the ocean of the order 0.6 PgC/yr globally, which should be included in any assessment of net air-sea and atmosphere-terrestrial biosphere transfers and ocean transports.

Uptake of Anthropogenic CO_2

Despite the importance of biological processes for the ocean's natural carbon cycle, current thinking maintains that the oceanic uptake of anthropogenic CO_2 is primarily a physically and chemically controlled process superimposed on a biologically driven carbon cycle that is close to steady state. This differs from the situation on land because of the different factors which control marine and terrestrial primary productivity. On land, experiments have repeatedly shown that current CO_2 concentrations are limiting to plant growth.

In the ocean, experimental evidence is against control of productivity by CO_2 concentrations, except for certain species at lower than contemporary CO_2 concentrations. Further, deep ocean concentrations of major nutrients and DIC are tightly correlated, with the existing ratios closely matching the nutritional requirements of marine organisms. This implies that as long as nutrients that are mixed into the ocean surface layer are largely removed by organic carbon production and export, then there is little potential to drive a net air-sea carbon transfer simply through alteration of the global rate of production.

Terrestrial ecosystems show greater variability in this respect because land plants have multiple ways to acquire nutrients, and have greater plasticity in their chemical composition. There are, however, extensive regions of the ocean surface where major nutrients are not fully depleted, and changes in these regions may play a significant role in altering atmosphereocean carbon partitioning. The increase of atmospheric pCO_2 over pre-industrial levels has tended to increase uptake into natural CO_2 sink regions and decreased release from natural outgassing regions.

Contemporary net air-sea fluxes comprise spatially-varying mixtures of natural and anthropogenic CO_2 flux components and cannot be equated with anthropogenic CO_2 uptake, except on a global scale. Uptake of anthropogenic CO_2 is strongest in regions where "old" waters, which have spent many years in the ocean interior since their last contact with the atmosphere, are re-exposed at the sea surface to a contemporary atmosphere which now contains anthropogenic CO_2. In an upwelling region, for example, the natural component of the air-sea flux may be to outgas CO2 to the atmosphere.

The higher atmospheric pCO_2 of the contemporary atmosphere acts to reduce this outgassing relative to the natural state, implying that more carbon remains in the ocean. This represents uptake of anthropogenic CO_2 by a region which is a source of CO_2 to the atmosphere. The additional carbon in the ocean resulting from such uptake is then transported by the surface ocean circulation, and eventually stored as surface waters sink, or are mixed, into the deep ocean interior.

Whereas upwelling into the surface layer is quantitatively balanced on a global scale by sinking, the locations where deep waters rise and sink can be separated by large horizontal distances.

Air-sea gas transfer allows older waters to approach a new steady state with higher atmospheric CO_2 levels after about a year at the sea surface. This is fast relative to the rate of ocean mixing, implying that anthropogenic CO_2 uptake is limited by the rate at which "older" waters are mixed towards the air-sea interface.

The rate of exposure of older, deeper waters is therefore a critical factor limiting the uptake of anthropogenic CO_2. In principle, there is sufficient uptake capacity in the ocean to incorporate 70 to 80% of anthropogenic CO_2 emissions to the atmosphere, even when total emissions of up to 4,500 PgC are considered. The finite rate of ocean mixing, however, means that it takes several hundred years to access this capacity.

Chemical neutralisation of added CO_2 through reaction with $CaCO_3$ contained in deep ocean sediments could potentially absorb a further 9 to 15% of the total emitted amount, reducing the airborne fraction of cumulative emissions by about a factor of 2; however the response time of deep ocean sediments is in the order of 5,000 years.

Using time-series and global survey data, the increasing oceanic carbon content has been directly observed, although the signal is small compared to natural variability and requires extremely accurate measurements. A longterm increase of surface water CO_2 levels tracking the mean atmospheric CO2 increase has been observed in the ocean's subtropical gyres and the equatorial Pacific.

However, very few such time-series exist and the response of other important oceanic regions to the atmospheric pCO_2 increase cannot yet be assessed. Inter-decadal increases in DIC concentrations at depth have been resolved from direct measurements. The total amounts of anthropogenic CO_2

accumulated in the ocean since the preindustrial era can also be estimated from measurements using recent refinements of long-standing methods for separating the natural and anthropogenic components of oceanic DIC.

FUTURE CHANGES IN OCEAN CO_2 UPTAKE

This section lists processes that may be important for the future uptake of anthropogenic CO_2. These changes can represent changes in anthropogenic CO_2 uptake itself (mainly physical and chemical processes), or changes in the natural biologically-linked cycling of carbon between the atmosphere and ocean.

Physical and Chemical Processes

Buffering Changes

The capacity of surface waters to take up anthropogenic CO_2 is decreasing as CO_2 levels increase. The magnitude of this effect is substantial. This decrease in uptake capacity of the ocean makes atmospheric CO_2 more sensitive to anthropogenic emissions and other changes in the natural cycling of carbon.

Emissions Rate

Even assuming no other changes to the carbon cycle, the proportion of emitted CO_2 that can be taken up by the ocean decreases as the rate of emission increases. This is due to the finite rate of exposure of 'older', deeper waters to the anthropogenic CO_2 contained in the atmosphere.

Warming

CO_2 is less soluble in warmer water, and the equilibrium pCO_2 in seawater increases by about 10 to 20 ppm per °C temperature increase. Warming of surface water would therefore tend to increase surface water pCO_2, driving CO_2 from the surface ocean to the atmosphere. The expected effect of such warming on atmospheric CO_2 may be smaller, depending on the rate of exchange between ocean surface waters and the deep ocean at high latitudes.

Vertical Mixing and Stratification

Several coupled atmosphere-ocean models have shown global warming to be accompanied by an increase in vertical stratification. Such a change would reduce the rate of mixing between surface and deep waters, and therefore reduce the effective volume of the ocean that is exposed to high atmospheric CO_2. On its own, this effect would tend to reduce the ocean CO_2 uptake. However, changes in stratification may also drive changes in the natural carbon cycle. The magnitude and even the sign of changes in the natural cycle are much more difficult to predict because of the complexity of ocean biological processes.

Biologically-linked Processes

Qualitative and quantitative changes in carbon uptake arising from changes in marine ecosystems are more speculative, but are likely to have occurred over glacial-interglacial time-scales. Falkowski *et al.* listed three major classes of biologically linked factors that can in principal alter the air-sea partitioning of CO_2:

- Changes in surface nutrient utilisation (e.g., in HNLC areas);
- Changes in total ocean content of major nutrients;
- Changes in the elemental composition of biogenic material (including the rain ratio).

Our incomplete understanding of present day nutrient controls on productivity limits our ability to predict future changes in ocean biology and their effect on CO_2 levels. For example, the possible identification of changes in deep ocean C:N:P ratios leaves open the question of the extent to which ocean biological carbon cycling is in steady state, or is likely to remain so in the future.

Changes in Surface Nutrient Utilisation

Changes in the utilisation of surface nutrients in HNLC

regions have the potential to alter export production and carbon storage in the ocean interior. Most attention focuses on the role of inadvertent or deliberate changes in the external supply of iron to such regions. The sign of possible future responses of ocean biota due to iron supply changes is difficult to assess.

Future iron supply may increase due to erosion (enhanced by agriculture and urbanisation) which tends to increase dust export and aeolian iron deposition. Conversely, a globally enhanced hydrological cycle and increased water-use efficiency of terrestrial plants may tend to reduce future dust export. The delivery of dust to the HNLC regions will be sensitive to regional changes in erosion and the hydrological cycle, affecting the important regions of dust export, rather than to global scale changes.

Surface nutrient supply could be reduced if ocean stratification reduces the supply of major nutrients carried to the surface waters from the deep ocean. The impact of stratification on marine productivity depends on the limiting factor. In regions limited by deep ocean nutrients, stratification would reduce marine productivity and the strength of the export of carbon by biological processes. Conversely, stratification also increases the light exposure of marine organisms, which would increase productivity in regions where light is limiting.

Changes in Total Ocean Content of Major Nutrients

Changes in the delivery of the major biologically limiting nutrients (N, P, Fe, Si) from riverine, atmospheric or sedimentary sources, or changes in removal rates (e.g., denitrification), could alter oceanic nutrient inventories and hence export production and ocean carbon storage. On the global scale, the upward fluxes of major nutrients are slightly depleted in N relative to P with respect to the nutrient requirements of phytoplankton.

This relative supply of N versus P may be sensitive to climate and circulation related changes in the rate of fixed-nitrogen removal by denitrification or via changes in the rate of nitrogen fixation. Changes in river flow and composition are also affecting the supply of nutrients.

The hypothesised link between nitrogen fixation in certain ocean regions and the external iron supply could play a role in future nutrient and carbon budgets. Nitrogen fixation rates may also be affected by changes in stratification and mixing. For example, Karl *et al.* have identified interannual variability in nitrogen fixation rates in the subtropical Pacific which are apparently linked to ENSO variability in upper ocean dynamics.

Changes in the Elemental Composition of Biogenic Material

The structure and biogeochemistry of marine ecosystems can be affected by numerous climate-related factors including temperature, cloudiness, nutrient availability, mixed-layer physics and sea-ice extent. In turn the structure of marine ecosystems, and particularly the species composition of phytoplankton, exert a control on the partitioning of carbon between the ocean and the atmosphere.

For example, a change in distribution of calcareous versus siliceous planktonic organisms could affect CO_2 uptake in the future, as it may have done in the past. Precipitation of $CaCO_3$ by marine organisms (calcification) removes dissolved CO_3, thus decreasing surface water alkalinity and reducing the capacity of sea water to dissolve atmospheric CO_2. Recent experimental evidence suggests that as a direct result of increasing atmospheric and surface water pCO_2 levels, oceanic calcification will decrease significantly over the next 100 years.

Model-based calculations suggest that decreases in coral reef calcification rates of the order 17 to 35% relative to pre-industrial rates are possible. Experimental studies with corals have confirmed such effects. Field and laboratory studies have shown that planktonic calcification is also highly sensitive to pCO_2 levels.

The calcification rate of coccolithophorids decreases by 16 to 83% at pCO_2 levels of 750 ppm. Such an effect would tend to favour CO_2 storage in the upper ocean and act as a negative feedback on atmospheric growth rates of CO_2. However, long-term predictions of such biological responses are hampered by a lack of understanding concerning physiological acclimation and genetic adaptations of species to increasing pCO_2.

PALAEO CO_2 AND NATURAL CHANGES IN THE CARBON CYCLE

Geological History of Atmospheric CO_2

Atmospheric CO_2 concentration has varied on all time-scales during the Earth's history. There is evidence for very high CO_2 concentrations (>3,000 ppm) between 600 and 400 Myr BP and between 200 and 150 Myr BP. On long time-scales, atmospheric CO_2 content is determined by the balancc among geochemical processes including organic carbon burial in sediments, silicate rock weathering, and vulcanism. In particular, terrestrial vegetation has enhanced the rate of silicate weathering, which consumes CO_2 while releasing base cations that end up in the ocean.

Subsequent deep-sea burial of Ca and Mg (as carbonates, for example $CaCO_3$) in the shells of marine organisms removes CO_2. The net effect of slight imbalances in the carbon cycle over tens to hundreds of millions of years has been to reduce atmospheric CO_2. The rates of these processes are extremely slow, hence they are of limited relevance to the atmospheric CO_2 response to emissions over the next hundred years.

It is pertinent, however, that photosynthesis evolved at a time when O_2 concentrations were far less than at present. O_2 has accumulated in the atmosphere over geological time because photosynthesis results in the burial of reduced chemical species: pyrite (FeS_2) derived from sulphur-reducing bacteria, and organic carbon.

This accumulation has consequences for terrestrial and marine ecosystems today. Primary production is carbon limited in terrestrial ecosystems in part because of (geologically speaking) low CO_2 concentrations, and in part because Rubisco (the enzyme that fixes CO_2 in all plants) also has an affinity for O_2 that reduces its efficiency in photosynthesis.

Primary production is iron limited in some marine ecosystems mainly because of the extreme insolubility of Fe(III), the predominant form of iron in the present, O_2-rich environment.

These difficulties faced by contemporary organisms represent a legacy of earlier evolution under very different biogeochemical conditions. In more recent times, atmospheric CO_2 concentration continued to fall after about 60 Myr BP and there is geochemical evidence that concentrations were <300 ppm by about 20 Myr BP.

Low CO_2 concentrations may have been the stimulus that favoured the evolution of C_4 plants, which increased greatly in abundance between 7 and 5 Myr BP. Although contemporary CO_2 concentrations were exceeded during earlier geological epochs, they are likely higher now than at any time during the past 20 million years.

Variations in Atmospheric CO_2 during Glacial/interglacial Cycles

The purity of Antarctic ice allows the CO_2 concentration in trapped air bubbles to be accurately measured. The CO_2 record from the Vostok ice core is the best available for the glacial/inter-glacial time-scale and covers the past four glacial/inter-glacial cycles (420 kyr) with a resolution of 1 to 2 kyr. The general pattern is clear: atmospheric CO_2 has been low (but 180 ppm) during glacial periods, and higher (but £300 ppm) during interglacials.

Natural processes during the glacialinterglacial cycles have maintained CO_2 concentrations within these bounds, despite considerable variability on multi-millenial time-scales. The present CO_2 concentration is higher than at any time during the 420 kyr period covered by the Vostok record. The terrestrial biosphere stores 300 to 700 Pg *more* carbon during interglacial periods than during glacial periods, based on a widely accepted interpretation of the ^{13}C record in deep-sea sediments.

Terrestrial modelling studies have reached the same conclusion. Thus, the terrestrial biosphere does not cause the difference in atmospheric CO_2 between glacial and interglacial periods. The cause must lie in the ocean, and indeed the amount of atmospheric change to be accounted for must be augmented to account for a fraction of the carbon transferred between the land and ocean.

The mechanism remains controversial. In part this is because a variety of processes that could be effective in altering CO_2 levels on a century time-scale can be largely cancelled on multi-millenial time-scales by changes in $CaCO_3$ sedimentation or dissolution. Orbital variations are the pacemaker of climate change on multi-millenial time-scales.

Atmospheric CO_2 is one of many Earth system variables that show the characteristic "Milankovitch" periodicities, and has been implicated as a key factor in locking natural climate changes to the 100 kyr eccentricity cycle. Whatever the mechanisms involved, lags of up to 2,000 to 4,000 years in the drawdown of CO_2 at the start of glacial periods suggests that the low CO_2 concentrations during glacial periods amplify the climate change but do not initiate glaciations.

Once established, the low CO_2 concentration is likely to have enhanced global cooling. During the last deglaciation, rising CO_2 paralleled Southern Hemisphere warming and was ahead of Northern Hemisphere warming. During glacial periods, the atmospheric CO_2 concentration does not track the "fast" changes in climate (e.g., decade to century scale warming events) associated with Dansgaard- Oeschger events, although there are CO_2 fluctuations of up to 20 ppm associated with the longer-lived events.

During the last deglaciation, atmospheric CO_2 concentration continued to increase, by about 12 ppm, through the Younger Dryas cold reversal seen in Northern Hemisphere palaeoclimate records. Palaeo-oceanographic evidence shows that the Younger Dryas event was marked by a prolonged shut-down of the thermohaline circulation, which is likely to have been triggered by the release of melt water into the North Atlantic.

Similar behaviour, with a slight rise in CO_2 accompanying a major Northern Hemisphere cooling and shutdown of North Atlantic Deep Water production, has been produced in a coupled atmosphere-ocean model. The observed CO_2 rise during the Younger Dryas period was modest, suggesting that atmospheric CO_2 has, under natural conditions, been well buffered against abrupt changes in climate, including thermohaline collapse. This buffering is a direct consequence of the large reservoir of DIC in the ocean.

Variations in Atmospheric CO_2 during the Past 11,000 Years

Natural variations in CO_2 during the past 11,000 years have been small (about 20 ppm) according to the best available measurements, which are from the Taylor Dome ice core. These measurements show a short-lived maximum around 11 kyr BP, followed by a slight fall, which may have been caused by increasing carbon storage in the terrestrial biosphere.

Atmospheric CO_2 concentration was about 260 ppm at its Holocene minimum around 8 kyr BP and increased towards about 280 ppm in the pre-industrial period. The same pattern and the same CO_2 concentration levels over the past 8 kyr have also been shown in another ice core, BH7 near Vostok. The causes of these changes are not known.

Preliminary ^{13}C measurements suggest that this increase may have been due to a gradual reduction in terrestrial carbon storage but others have considered an oceanic explanation more likely. Atmospheric CO_2 concentrations have also been reconstructed indirectly, from stomatal index measurements on sub-fossil leaves.

Stomatal density and stomatal index of many species respond to atmospheric CO_2 but are influenced by other environmental variables as well.

One recent stomatal index record, interpreted as implying high (up to 350 ppm) and rapidly fluctuating CO_2 concentrations in the early Holocene, is clearly incompatible with the ice core record of Indermühle *et al.*, whereas a continuous stomatal index record from 9 kyr BP onwards has shown concentration trends consistent with the ice-core records.

The excellent agreement among different high-resolution Antarctic ice cores covering the past 1,000 years. Atmospheric CO_2 concentration fell by about 8 to 10 ppm during the Little Ice Age. A slight contemporaneous increase in ^{13}C of atmospheric CO_2 has led to the suggestion that this effect was caused by enhanced carbon storage on land.

Implications

The Vostok record of atmospheric CO_2 and Antarctic climate is consistent with a view of the climate system in which CO_2 concentration changes amplify orbitally-induced climate changes on glacial/inter-glacial time-scales. Changes during the present inter-glacial (until the start of the anthropogenic CO_2 rise) have been small by comparison. Although complete explanations for these changes in the past are lacking, high-resolution ice core records establish that the human-induced increase of atmospheric CO_2 over the past century is at least an order of magnitude faster than has occurred during the preceeding 20,000 years.

Anthropogenic Sources of CO_2

Emissions from Fossil Fuel Burning and Cement Production

Current anthropogenic emissions of CO_2 are primarily the result of the consumption of energy from fossil fuels. Estimates of annual global emissions from fossil fuel burning and cement production have been made for the period from 1751 through 1999. Estimates of annual global emissions from fossil fuel burning and cement production by Marland *et al.* span the period from 1751 through to 1997, reaching a maximum in 1997 of 6.6 PgC/yr (0.2 PgC/yr of this was from cement production). The primary data for these estimates are annual energy statistics compiled by the United Nations. Emissions for 1998 and 1999 have been estimated based on energy statistics compiled by British-Petroleum. Emission factors were applied to consumption statistics to calculate emissions over the period 1990 to 1999.

Emissions were then scaled to match the estimates for emissions from fossil fuel burning and cement production from Marland *et al.* over the overlap period from 1990 to 1997. The scaled emission estimates, therefore, implicitly include emissions from cement production. The average value of emissions for the 1980s given by Marland *et al.* is 5.44 ± 0.3 PgC/yr, revised from the earlier estimate of 5.46 ± 0.3 PgC/yr used in the SAR and in the Special Report on Radiative Forcing (hereafter SRRF). Estimated emissions rose from 6.1 PgC/yr in 1990 to 6.5 PgC/yr in 1999. The average value of emissions in the 1990s was 6.3 ± 0.4 PgC/yr.

Consequences of Land-use Change

About 10 to 30% of the current total anthropogenic emissions of CO_2 are estimated to be caused by land-use conversion. Such estimates rely on land cover data sets which are highly variable, and estimates of average carbon density of vegetation types, which are also highly variable with stand age and local conditions. Hence they cannot be specified as accurately as is possible for fossil fuel emissions.

Net land-use flux, comprising the balance of positive terms due to deforestation and negative terms due to regrowth on abandoned agricultural land, has been estimated based on landuse statistics and simple models of rates of decomposition and regrowth, excluding possible climate, CO_2 and N fertilisation effects.

Not all land-use emissions are included, for example mining of peatlands. The analysis of Houghton indicated that the net flux due to land-use change was 2.0 ± 0.8 PgC/yr during the 1980s, almost entirely due to deforestation of tropical regions. Temperate forests were found to show an approximate balance between carbon uptake in regrowing forests and carbon lost in oxidation of wood products, except in Europe, which showed a small net accumulation.

The estimate of 2.0 PgC/yr is somewhat higher than Houghton and Hackler's earlier estimate of 1.6 PgC/yr for the same period, which was used in the SAR, because of a reanalysis of data from tropical Asia.

However, other recent analyses by the same authors reduce the estimated emissions from the Brazilian Amazon by half, and point to other previously unaccounted for sinks of carbon in the USA such as fire suppression and woody encroachment, and changes in the management of agricultural soils.

Consideration of these additional studies brings the overall total back down to 1.7 ± 0.8 PgC/yr, as given in the SRLULUCF. An independent analysis by the Carbon Cycle Model Linkage Project (CCMLP) also calculated the marginal effects of land-use changes on the global terrestrial carbon budget. Land-use change data (conversions between native vegetation and crops) were derived from Ramankutty and Foley.

The estimates obtained for net land-use flux during the 1980s were between 0.6 and 1.0 PgC/yr, i.e., substantially smaller than the fluxes calculated by Houghton. The reasons for this discrepancy are unclear. The CCMLP estimates may be too low because they neglected conversions to pasture. However, data presented in Houghton indicate that the main changes during recent decades were due to land conversion for crops.

A more important difference may lie in the timing of deforestation in different regions in the tropics, where Ramankutty and Foley show higher overall rates in the 1970s and lower rates in the 1980s than Houghton does. Another analysis calculated a substantially higher net source due to land-use change in the tropics of 2.4 ± 1.0 PgC/yr during the 1980s. This analysis did not deal with temperate regions, and is not used in the global budget estimates. No complete global assessment of deforestation effects covering the 1990s is available. Rates of deforestation appear to be declining.

The FAO tropical forest assessment reported annual losses of 15.5×10^6 ha in the 1980s, and 13.7×10^6 ha in 1990 to 1995. Independent studies show a significant decline in deforestation rates in the Amazon region. The annual flux of carbon from land-use change for the period from 1990 to 1995 has been estimated to be 1.6 PgC/yr from 1990 to 1995, consisting of a source of 1.7 PgC/yr in the tropics and a small sink in temperate and boreal areas.

Atmospheric Measurements and Global CO_2 Budgets

Continuous time-series of highly precise measurements of the atmospheric composition are of central importance to current understanding of the contemporary carbon cycle. CO_2 has been measured at the Mauna Loa and South Pole stations since 1957, and through a global surface sampling network developed in the 1970s that is becoming progressively more extensive and better inter-calibrated. Associated measurements of 13_C in atmospheric CO_2 began in 1977.

More recently, complementary information has been available from O_2 concentrations, which have been regularly measured since the early 1990s. O_2 concentration data for the 1980s have been gleaned by two methods: sampling of archived air flasks that were collected during the 1980s, and measuring the air trapped in Antarctic firn. In addition to fossil fuel CO_2 emissions, the observed seasonally corrected growth rate of the atmospheric CO_2 concentrations, based on the two longest running atmospheric CO_2 recording stations.

It is evident from this comparison that a part of the anthropogenic CO_2 has not remained in the atmosphere; in other words, CO_2 has been taken up by the land or the ocean or both. This comparison also shows that there is considerable interannual variability in the total rate of uptake. O_2 and CO_2 measurements are used here to provide observationally- based budgets of atmospheric CO_2.

CO_2 budgets are presented here for the 1980s, and for the 1990s. The reported error ranges are based on uncertainties of global fossil fuel emissions, determination of the decadal average changes in the atmospheric CO_2 concentration, and O_2:N_2 ratio; and uncertainties in the assumed O_2:CO_2 stoichiometric ratios in the combustion of fossil fuels and in photosynthesis and respiration.

The error ranges reflect *uncertainties* of the decadal mean averaged values; they do not reflect interannual variability in annual values, which far exceeds uncertainty in the decadal mean rate of increase.

The salient facts are as follows:

- During the 1980s, fossil fuel emissions were on average 5.4 ± 0.3 PgC/yr and atmospheric CO_2 content increased on average by 3.3 ± 0.1 PgC/yr. Partitioning of CO_2 uptake was estimated based on archived flask O_2 measurements for the 1979 to 1997 period, taking the O_2 trend during 1991 to 1997 into account. The resulting estimate of the ocean-atmosphere flux was -1.9 ± 0.6 PgC/yr and of the landatmosphere flux -0.2 ± 0.7 PgC/yr. It is corroborated by independent O2 measurements in Antarctic firn. Restricting the analysis to the Battle *et al.* data for the 1980 to 1989 period, an ocean-atmosphere flux of -1.8 ± 1.0 PgC/yr and a land-atmosphere flux of -0.4 ± 1.0 PgC/yr were obtained.
- Despite a greater emission rate of 6.3 ± 0.4 PgC/yr, the average atmospheric increase during the 1990s was 3.2 ± 0.1 PgC/yr, i.e., about the same as during the 1980s. An exceptionally low rate of increase during the early 1990s was balanced by a high rate during the late 1990s. Based on the longest existing O_2 records from La Jolla and Alert (northern Canada), the ocean-atmosphere flux during the 1990s was -1.7 ± 0.5 PgC/yr and the land-atmosphere flux was -1.4 ± 0.7 PgC/yr. Ocean uptake in the 1980s as estimated from O_2 and CO_2 measurements thus agrees with the estimates in the SRRF and the SAR (although these were model-based estimates; this section presents only observationallybased estimates). Considering the uncertainties, the ocean sink in the 1990s was not significantly different from that in the 1980s. The land-atmosphere flux was close to zero in the 1980s, as also implied by the SAR budget. The land appears to have taken up more carbon during the 1990s than

during the 1980s. The causes cannot yet be reliably quantified, but possible mechanisms include a slow down in deforestation, and climate variability that resulted in temporarily increased land and/or ocean uptake in the early 1990s. These budgets are consistent with information from atmospheric ^{13}C measurements and with budgets presented in the SRLULUCF except that estimated ocean uptake is smaller, and land uptake accordingly larger, than given in the SRLULUCF. Several alternative approaches to estimating the oceanatmosphere and land-atmosphere fluxes of CO_2. Alternative methods for estimating the global oceanatmosphere flux, based on surface-water pCO_2 measurements and ocean ^{13}C changes, respectively, have yielded a range of -1.5 to -2.8 PgC/yr (for various recent periods). The total anthropogenic CO_2 added to the ocean since pre-industrial times can also be estimated indirectly using oceanic observations. A global value of 107 ± 27 PgC by 1990 can be estimated from the basin-scale values of 40 ± 9 PgC for the Atlantic in the 1980s, 20 ± 3 PgC for the Indian Ocean in 1995, and the preliminary value of 45 PgC for the Pacific Ocean in 1990 to 1996 with a large uncertainty of the order of ± 15 PgC. Assuming that accumulation of CO_2 in the ocean follows a curve similar to the (better known) accumulation in the atmosphere, the value for the ocean-atmosphere flux for 1980 to 1989 would be between -1.6 and -2.7 PgC/yr. The land-atmosphere flux based on atmospheric measurements represents the *balance* of a net land-use flux (currently a positive flux, or carbon source, dominated by tropical deforestation) and a residual component which is, by inference, a negative flux or carbon sink. Using the land-atmosphere flux estimates, assuming that land-use change contributed +1.7 PgC/yr to the atmosphere during the 1980s, then a residual terrestrial flux of -1.9 PgC/yr (i.e., a residual sink of similar magnitude to the total ocean uptake) is required for mass balance. This is the term popularly (and misleadingly) known as the "missing sink". The central estimate of its magnitude agrees with previous analyses, e.g., in the SAR (if "northern forest regrowth" is combined with "residual terrestrial sink" terms in the SAR budget) and the SRLULUCF. The uncertainty around this number is rather large, however, because it compounds the uncertainty in the atmospheric budget with a major uncertainty about changes in land use. Using an error range corresponding to 90% confidence intervals around the atmospheric estimate of -0.2 PgC/yr (i.e., 1.6 ó, giving confidence intervals of ±1.1 PgC/yr), and taking the range of estimates for CO_2 released due to land-use change during the 1980s, the residual terrestrial sink is estimated to range from -3.8 to +0.3 PgC/yr for the 1980s. Model-based analysis of the components of the residual terrestrial sink.

Interannual Variability in the Rate of Atmospheric CO_2 Increase

The rate of increase in the globally averaged atmospheric concentration of CO_2 varies greatly from year to year. "Fast" and "slow" years have differed by 3 to 4 PgC/yr within a decade. This variability cannot be accounted for by fossil fuel emissions, which do not show short-term variability of this magnitude. The explanation must lie in variability of the land-atmosphere flux or the ocean-atmosphere flux or both. Variability in both systems could be induced by climate variability.

An association between CO_2 variability and El Niño in particular has been reported for over twenty years and has been confirmed by recent statistical analyses. During most of the observational record, El Niño events have been marked by high rates of increase in atmospheric CO_2 concentration compared with surrounding years, in the order of > 1 PgC/yr higher during most El Niño events. Direct measurements of oceanic CO_2 in the equatorial Pacific over the last 20 years have shown that the natural efflux of CO_2 from this region is reduced by between 0.2 to 1.0 PgC/yr during El Niño, mainly due to the reduced upwelling of CO_2-rich waters.

The ocean response to El Niño in the most active region thus tends to increase global CO_2 uptake, counter to the increasing atmospheric concentration. Although it cannot be ruled out that other ocean basins may play a significant role for global interannual variability in ocean-atmosphere flux, the existing oceanic measurements suggest (by default) that the response of the terrestrial biosphere is the cause of the typically high rates of CO_2 increase during El Niño. Associated variations in the north-south gradient of CO_2 indicate that the El Niño CO_2 anomalies originate in the tropics.

Typical El Niño events are characterised by changed atmospheric circulation and precipitation patterns that give rise to high tropical land temperatures (which would be expected to increase Rh and reduce NPP); concurrent droughts which reduce NPP, especially in the most productive regions such as the Amazon rain forest; and increased incidence of fires in tropical regions.

Increased cloudiness associated with enhanced south-east Asian monsoons during the late phase of El Niño has also been suggested as a factor reducing global NPP. Typically, although not invariably, the rate of atmospheric CO_2 increase declines around the start of an El Niño, then rapidly rises during the late stages.

It has been suggested that this pattern represents early onset of enhanced ocean CO_2 uptake, followed by reduced terrestrial CO_2 uptake or terrestrial CO_2 release. Atmospheric ^{13}C and, more recently, O_2 measurements have been used to partition the interannual variability of the atmospheric CO2 increase into oceanic and terrestrial components. Analyses based on ^{13}C by Keeling *et al.* and Francey *et al.* reached contradictory conclusions, but the discrepancies are now thought to be due at least in part to ^{13}C measurement calibration problems during the 1980s, which have largely been resolved during the 1990s.

For the 1990s, a range of analyses using different atmospheric observations and/or data analysis techniques estimate that the amplitude of annual peak to peak variation associated with the ocean is about 2 to 3 PgC/yr and the amplitude associated with the terrestrial biosphere is about 4 to 5 PgC/yr. A similar partitioning was estimated by Bousquet *et al.* based on the spatial pattern of CO_2 measurements. However, the various reconstructed time sequences of terrestrial and ocean uptake differ in many details and do not provide conclusive evidence of the mechanisms involved.

The early 1990s were unusual in that the growth rate in atmospheric CO_2 was low (1.9 PgC/yr), especially in the Northern Hemisphere, while an extended El Niño event occurred in the equatorial Pacific. Various mechanisms have been suggested, but none fully explain this unusual behaviour of the carbon cycle.

The slow down in the CO_2 increase has been linked to the predominantly mid- to high latitude cooling caused by the Pinatubo eruption, but there is no proof of any connection between these events. Other partial explanations could come from a temporary slow down of tropical deforestation, or natural decadal variability in the ocean-atmosphere or land-atmosphere fluxes. In any case, the slowdown proved to be temporary, and the El Niño of 1998 was marked by the highest rate of CO_2 increase on record, 6.0 PgC/yr.

Inverse Modelling of Carbon Sources and Sinks

Inverse modelling attempts to resolve regional patterns of CO_2 uptake and release from observed spatial and temporal patterns in atmospheric CO_2 concentrations, sometimes also taking into consideration O_2 and/or ^{13}C measurements. The most robust results are for the latitudinal partitioning of sources and sinks between northern and southern mid- to high latitudes and the tropics. The observed annual mean latitudinal gradient of atmospheric CO_2 concentration during the last 20 years is relatively large (about 3 to 4 ppm) compared with current measurement accuracy.

It is ... ver not as large as would be predicted from the geographical distribution of fossil fuel burning – a fact that suggests the existence of a northern sink for CO_2, as already recognised a decade ago. The nature of this sink, however, cannot be determined from atmospheric CO_2 concentration measurements alone. It might reflect, at least in part, a natural source-sink pattern of oceanic CO_2 fluxes.

This view is supported by the early atmospheric CO_2 data from the 1960s which do not show a clear latitudinal gradient, despite the fact that at that time the fossil emissions were already at least half as large as in the 1990s. Quantitative analysis shows that the Northern Hemisphere sink has not changed much in magnitude since the 1960s.

On the other hand, the existing airsea flux measurements do not support the idea of a large oceanic uptake of CO_2 in the Northern Hemisphere. An alternative view, therefore, locates a significant fraction of this Northern Hemisphere sink on land.

This view is corroborated, at least for the 1990s, by analyses of the concurrent latitudinal gradients of ^{13}C. Results of analyses for the 1980s and 1990 to 1996, carried out by eight modelling groups using different atmospheric transport models, observational data, constraints and mathematical procedures.

Only the most robust findings, i.e., estimates of the mean carbon balance for three latitude bands averaged over the two time periods, are shown. The latitude bands are: "southern extratropics" (>30°S), "tropics" (30°S to 30°N) and "northern extratropics" (>30°N).

The carbon balance estimates are broken down into land and ocean compartments within each latitude band. The inferred ocean uptake pattern shows the sum of two components: the natural carbon cycle in which CO_2 is outgassed in the tropics and taken up in the extratropics, and the perturbation uptake of anthropogenic CO_2.

Separation of these two components cannot be achieved from atmospheric measurements alone. The estimates for the land, on the other hand, in principle indicate the locations of terrestrial anthropogenic CO_2 uptake (albeit with caveats listed below).

For 1980 to 1989, the inversemodel estimates of the land-atmosphere flux are -2.3 to -0.6 PgC/yr in the northern extratropics and -1.0 to +1.5 PgC/yr in the tropics. These estimates imply that anthropogenic CO_2 was taken up *both* in the northern extratropics and in the tropics (balancing deforestation). The estimated landatmosphere flux in the southern extratropics is estimated as close to zero, which is expected given the small land area involved. Estimates of CO_2 fluxes for the period 1990 to 1996 show a general resemblance to those for the 1980s. For 1990 to 1996, the inverse-model estimates of the land-atmosphere flux are -1.8 to -0.7 PgC/yr in the northern extratropics and -1.3 to +1.1 PgC/yr in the tropics.

These results suggest a tendency towards a reduced land-atmosphere flux in the tropics, compared to the 1980s. Such a trend could be produced by reduced deforestation, increased CO_2 uptake or a combination of these. Inverse modelling studies usually attempt greater spatial resolution of sources and sinks than is presented in this section. However, there are large unresolved differences in longitudinal patterns obtained by inverse modelling, especially in the northern hemisphere and in the tropics. These differences may be traced to different approaches and several difficulties in inverse modelling of atmospheric CO_2.

Terrestrial Biomass Inventories

Inventory studies measure changes in carbon stocks over large areas, and can thus provide spatially aggregated estimates of large-scale fluxes of CO_2 over multi-annual time-scales. Mid- and high latitude forests are covered by extensive national inventories based on repeated measurements of many thousands of plots.

Inventories in the tropics are by comparison generally inadequate, particularly in view of the high rates of land-use change and extremely heterogeneous carbon density in many tropical ecosystems. There are still therefore large uncertainties in attempting to balance the terrestrial carbon budget on a global scale using inventory data.

The FAO Temperate and Boreal Forest Resource Assessment is a recent synthesis of inventories of forests and other wooded lands in Annex (developed) countries for the early 1990s. Many countries reported substantial increases in forest areas in recent years, as well as increasing carbon density in existing forests.

According to TBFRA-2000, the land-atmosphere flux was -0.9 PgC/yr for all Annex I countries combined (the net annual increment of trees accounted for -1.5 PgC/yr, while losses due to fellings were 0.6 PgC/yr). Of this flux, -0.8 PgC/yr was due to uptake in "northern" forests (Europe, CIS, Japan and North America). An earlier review of individual regional and national studies by Dixon *et al.*, highlighted in the IPCC WGII Second Assessment Report, gave a range of -0.6 to -0.9 PgC/yr for the land atmosphere flux in northern forests.

While TBFRA- 2000 estimated biomass of woody vegetation only, the analyses reviewed in Dixon *et al.* included other vegetation, soils, litter and wood products. Under the United Nations Framework Convention for Climate Change (UNFCCC) signatory countries are required to report greenhouse gas emissions, including those from land-use change and forestry.

Compilation of these data implies a land-atmosphere flux of -0.6 PgC/yr for all Annex I countries, and -0.6 PgC/yr for Annex I countries in the northern latitudes only. While the TBFRA synthesized country statistics and adjusted data to fit FAO definitions and methodologies for calculating carbon stocks, the UNFCCC report summarises emissions data reported by each country according to IPCC guidelines; interpretation of guidelines is variable, and not all countries had reported data on land use. The implications of definitions and methodologies in calculating carbon fluxes, particularly in relation to implementation of the Kyoto Protocol, is discussed in detail in the SRLULUCF.

A recent compilation of data from 478 permanent plots in mature tropical moist forests throughout the tropics over at least two decades found these were taking up carbon due to increasing rates of tree growth. Extrapolation from these plots led to an estimated land-atmosphere flux of (0.6 ± 0.3 PgC/yr in Latin America; growth trends in African and Asian forests were not significantly different from zero. This net uptake is offset by emissions due to deforestation.

Dixon *et al.* estimated tropical forests overall to be a net source of carbon with a land-atmosphere flux 1.7 ± 0.4, based mostly on FAO inventory data and simple models of the effect of land-use change.

It will not be possible to assess trends and fluxes for the 1990s in the tropics from inventory data until a full data set is available from the FAO Global Forest Resources Assessment 2000. Among those countries that have reported land-use emissions data to the UNFCCC, there are significant discrepancies between the primary data used in emissions inventories and the data available in international surveys; for example, rates of deforestation differ from rates reported by FAO by as much as a factor of six. The results of globally aggregated forest inventories show a greater uptake of carbon in forest growth than model-based calculations of the marginal effects of land-use change.

Thus, inventory studies provide independent evidence for the existence of a residual terrestrial sink; and they show that a substantial part of this sink, at least, is located in northern extratropical and tropical forests. Additional evidence from individual inventory studies in mature forests that have not undergone land-use changes shows that carbon stocks in such forests are increasing.

The difference between the northern extra-tropical land-atmosphere flux of around -0.8 PgC/yr calculated by inventories and that of - 0.1 PgC/yr from land-use statistics, both for the early 1990s, implies a residual terrestrial sink on the order of -0.7 PgC/yr in northern mid- and high latitudes. Combining this with the estimated sink of -0.6 PgC/yr in mature tropical moist forests makes it plausible that at least a significant fraction of the current global terrestrial sink could be explained by an increase of carbon stocks in extant forests.

The inventory-based estimate of landatmosphere flux in northern forests (-0.8 PgC/yr) is at the positive end of the range calculated by inverse modelling studies for the >30°N latitude band from 1990 to 1996 (-1.8 to -0.7 PgC/yr), either because of biases in inverse modelling that might tend to increase apparent uptake in the north, or because possible sinks in other ecosystems (e.g., temperate grassland soils) have not been considered in the inventories.

In the tropics, the difference between the uptake of carbon estimated by inventory studies in mature forests of Latin America (-0.6 PgC/yr) and the estimated emissions due to deforestation in the tropics of 1.7 PgC/yr yields an estimated landatmosphere flux of 1.1 PgC/yr, which is at the positive end of the range calculated by inverse modelling studies for 30°S to 30°N (-1.3 to +1.1 PgC/yr). Again, it should be noted that possible additional sinks (e.g., in savannas) are neglected by the land-use and inventory-based calculations.

CARBON CYCLE MODEL EVALUATION

Terrestrial and Ocean Biogeochemistry Models

The interactions of complex processes can be analysed with models that incorporate current knowledge at the process level, including syntheses of experimental results. Process-based models make it possible to explore the potential consequences of climate variability for the global carbon cycle, and to project possible future changes in carbon cycling associated with changes in atmospheric and ocean circulation.

Models can be run with prescribed inputs such as observations of surface climate and CO_2 or the output of climate models. They can also be coupled to atmospheric general circulation models, to allow simulation of a wider range of interactions between climate and the carbon cycle.

Process-based terrestrial models used in carbon cycle studies are:

- Terrestrial biogeochemical models (TBMs), which simulate fluxes of carbon, water and nitrogen coupled within terrestrial ecosystems,
- Dynamic global vegetation models (DGVMs), which

further couple these processes interactively with changes in ecosystem structure and composition (competition among different plant functional types).

The treatment of carbon-nutrient interaction varies widely; for example, some models treat nitrogen supply explicitly as a constraint on NPP, while others do not. There are currently about 30 TBMs and <10 DGVMs.

Cramer and Field and Cramer *et al.* reported results from intercomparisons of TBMs and DGVMs respectively.

A current international project, Ecosystem Model/Data Intercomparison (EMDI), aims to test models of both types against a large set of terrestrial measurements, in order to better constrain the modelled responses of terrestrial carbon cycling to changes in CO_2 and climate.

Process-based ocean models used in carbon cycle studies include surface exchange of CO_2 with the atmosphere, carbon chemistry, transport by physical processes in the ocean, and transport by marine biology.

The parametrization of marine biology can be classified as:

- Nutrient-based models where the export of carbon below the surface ocean (approximately the top 50 m) is a function of surface nutrient concentration,
- Nutrientrestoring models in which biological carbon fluxes are set to the rates required for maintaining observed nutrient concentration gradients against dissipation by ocean mixing,
- Models that explicitly represent the food chain involving nutrients, phytoplankton, zooplankton and detritus (NPZD models). In current models, the uptake of anthropogenic CO_2 is controlled mainly by physical transport and surface carbon chemistry, whereas the natural carbon cycle is controlled by physical, chemical and biological processes. The Ocean Carbon Cycle Model Intercomparison Project (OCMIP) compared the performance of four ocean models with respect to natural and anthropogenic tracers, and is currently undergoing a similar comparison with 13 models and an extended data set.

Evaluation of Terrestrial Models

Evaluation of terrestrial carbon cycle models requires different types of data to test processes operating on a range of time-scales from hours to centuries, including short-term environmental responses of CO_2 and water fluxes between vegetation canopies and the atmosphere, responses of ecosystem carbon balance to interannual climate variability, and longer-term consequences of historical land-use change. Differences and uncertainties in model behaviour have been evaluated through model intercomparison and sensitivity analyses.

Natural Carbon Cycling on Land

Terrestrial model evaluation has traditionally been carried out as comparisons with *in situ* field observations of ecosystem variables.

The largest data sets of relevant field measurements are for N.P.P. and soil carbon content. Other "target" variables include soil moisture, nitrogen mineralisation rate, and the amounts of carbon and nitrogen in different compartments of the ecosystem.

Such comparisons have generally shown reasonable agreement between observed and modelled geographic patterns of these variables, but they do not test the time-dependent response of models to environmental variability. Time-dependent data sets for *in situ* comparisons are now becoming available, thanks to eddy-covariance measurements of CO_2 fluxes.

Daily and seasonal cycles of CO_2 and water fluxes provide a test of the coupling between the carbon and hydrological cycles as simulated by terrestrial models.

Flux measurements are now being carried out on a multi-annual basis at an increasing number of stations, although global coverage remains uneven, with the greatest concentration in Europe and North America and few measurements from the tropics. Field campaigns have started to retrieve flux data from more remote regions. The Large-scale Biosphere Atmosphere Experiment in Amazonia, LBA, will yield more comprehensive data on the carbon, water and energy exchanges of tropical terrestrial ecosystems and will allow a more rigorous evaluation of the performance of models in the tropics than has been possible up until now.

As current models show conflicting responses of global NPP to climate, systematic comparisons with seasonal and interannual flux measurements are a priority to reduce uncertainties in terrestrial carbon modelling. Terrestrial models have also been evaluated at a global scale by comparing simulated ecosystem water balance with river runoff, and simulated seasonal leaf area with satellite observations of "greenness", often based on the normalised difference vegetation index (NDVI).

NDVI data can be translated into estimates of the plant-absorbed fraction of incoming photosynthetically active radiation (FPAR), which is related to leaf area index (LAI). The first terrestrial model intercomparison showed differences among model simulations of LAI and its seasonality. More recently, it has been shown that constraining a terrestrial model with remotely sensed spatial patterns of FPAR can lead to a reduction of uncertainty in NPP simulations by about one third. Agreement with patterns of remotely sensed FPAR has thus become a standard benchmark for terrestrial models and attention has been focused on improving the simulation of LAI and its seasonal variations.

A more direct test of the simulated net exchange of CO_2 between the terrestrial biosphere and the atmosphere is provided by comparison with atmospheric CO_2 measurements at remote monitoring sites. The comparison requires the use of an atmospheric transport model to simulate CO_2 as a passive tracer. The seasonal cycle of atmospheric CO_2 shows a strong latitudinal pattern in amplitude and phase, and is dominated by the terrestrial biosphere. The ability to simulate this seasonal cycle thus constitutes a benchmark for terrestrial models' response to climate. Generally, the observed seasonal cycles of CO_2 in northern and tropical latitudes can be well simulated, with terrestrial models using NDVI data as input, or by fully prognostic models, including DGVMs.

Major features of interannual variability of the CO_2 increase are also simulated by terrestrial models. This finding supports the hypothesis that terrestrial effects are important in determining the interannual variability of CO_2 uptake. During typical El Niño events, terrestrial model results consistently show strongly reduced CO_2 uptake or CO_2 release by the land.

This result has been obtained with a range of models, even when the models differ substantially in the relative sensitivitivities of NPP and heterotrophic respiration to temperature. The low CO_2 growth rate during the early 1990s has been simulated by some terrestrial models. At the longest time-scales of interest, spanning the industrial period, models of the natural terrestrial carbon cycle show a pronounced response to rising atmospheric CO_2 levels as a result of CO_2 fertilisation, generally larger than the NPP response to the climate change over this period. According to CCMLP results, the CO_2 increase maintains a lead of NPP over Rh and an increase of the amplitude of the seasonal CO_2 cycle, consistent with long-term observations, which indicate an increase in amplitude of about 20% since accurate atmospheric measurements began.

However, the magnitude of this effect was greatly over- or under estimated by some models, reflecting unresolved differences in the parametrization of the CO_2 fertilisation response.

Uptake and Release of Anthropogenic CO_2 by the Land

The most comprehensive model-based estimates of the terrestrial components of the anthropogenic CO_2 budget are those that have been produced by the CCMLP. McGuire *et al.* used two TBMs and two DGVMs driven by changes in atmospheric CO_2, then changes in CO_2 with historical changes in climate (from observations), and finally changes in CO_2 and climate with landuse change from Ramankutty and Foley. In these simulations, CO_2 fertilisation accounted for a landatmosphere flux of -0.9 to -3.1 PgC/yr, land-use change a positive flux of 0.6 to 1.0 PgC/yr, and climate variability a small additional effect of uncertain sign, -0.2 to 0.9 PgC/yr during the 1980s.

The total land-atmosphere flux simulated for the 1980s amounted to -0.3 to -1.5 PgC/yr, which is consistent with or slightly more negative than the observationally-based estimate of -0.2 ± 0.7 PgC/yr. Net uptake by all models reported in McGuire *et al.* is shown to be occurring mainly in tropical, temperate and boreal forests - consistent with forest inventory data– while some regions (notably semi-arid tropical and sub-tropical regions) show net carbon loss. The model estimates of the CO_2 source due to land-use change are substantially smaller than the estimate of Houghton.

This divergence primarily reflects disagreements between the Houghton and Ramankutty and Foley data sets as to the timing of tropical deforestation in different regions. There is no general agreement on how to model the linkage between reactive nitrogen deposition and vegetation productivity, and recent model estimates of the additional effect of anthropogenic nitrogen fertilisation on the global carbon cycle vary widely.

The anthropogenic nitrogen input itself, the fate of anthropogenic nitrogen in the ecosystem, and changes in ecosystem nitrogen fixation represent major sources of uncertainty. Estimates of the anthropogenic nitrogen effect range from -0.2 PgC/yr to -1.1 or -1.4 PgC/yr. The model with the smallest CO_2 fertilisation effect (-0.9 PgC/yr) in the McGuire *et al.* study has been shown to respond strongly to anthropogenic nitrogen input, yielding a combined (CO_2 and nitrogen) fertilisation effect of -1.5 PgC/yr.

A modelling study by Lloyd suggests that CO_2 and nitrogen fertilisation effects may by synergistic. Evaluation of model results on carbon-nitrogen coupling against experimental results is a current research focus.

Evaluation of Ocean Models

Natural and anthropogenic tracers have been extensively measured, most recently as part of the Joint Global Ocean Flux

Study (JGOFS) and World Ocean Circulation Experiment (WOCE). Because of these measurement campaigns, such tracers provide important opportunities to evaluate representations of ocean physics and biogeochemistry in models.

Natural Carbon Cycling in the Ocean

Most global ocean models of the carbon cycle are successful in reproducing the main vertical and horizontal features of ocean carbon content.

The observed features reasonably reproduced by all ocean models are the mean vertical gradient in DIC, with enriched deep ocean concentrations, and the spatial patterns of surface pCO_2 with outgassing in the tropics and uptake at higher latitudes.

Furthermore, models which incorporate marine biology (including DOC and plankton dynamics) roughly reproduce the seasonal cycle of surface ocean pCO_2, atmospheric O_2 after it has been corrected for seasonal land variability, and surface chlorophyll.

Ocean carbon models can also roughly reproduce the phase and amplitude of interannual variability of ocean pCO_2 in the equatorial Pacific in agreement with available observations.

Although many first-order features can be reproduced by global models, there are still important aspects of the ocean carbon cycle that are not well simulated, because either marine biology or ocean physics are imperfectly reproduced. Ocean carbon models have difficulties in reproducing the spatial structure of the deep ocean 14_C, which suggests problems in simulating the physical exchange of carbon between surface and the deep ocean.

Models display their largest disagreements where fewest observations exist, in particular in the important region of the Southern Ocean where the mixing of tracers is subject to large uncertainties.

In spite of these differences, all ocean carbon models estimate zero interhemispheric transport of carbon whereas a transport as large as 1 PgC/yr has been inferred from atmospheric CO_2 measurements. Consideration of the global transport of carbon by rivers reduces the discrepancy but does not remove it.

Atmospheric CO_2 and O_2 measurements suggest that interhemispheric transport may be incorrectly simulated by ocean models, and could hint at difficulties in modelling heat transport. Recent data from the Southern Ocean, however, seem closer to model results and the question about interhemispheric transport thus remains open. These problems could partly be resolved by a better representation of the physical transport of carbon in the ocean, especially isopycnal diffusion, sub-grid eddy mixing, and sea-ice formation.

Three common problems related to marine biology in global ocean models are discussed here. First, most models poorly represent the formation and dissolution of $CaCO_3$, which controls alkalinity. This process is often parameterized as a function of direct or indirect observations (salinity, temperature, nutrients).

Although correct for the present day ocean, this parametrisation may not hold for past or future conditions with different ocean circulation and surface water fluxes. The alkalinity cycle is difficult to represent because the rate of $CaCO_3$ formation derived from observations is consistently larger than the one required by models for reproducing observed deep ocean alkalinity.

Second, marine productivity tends to be underestimated by models in sub-tropical regions and overestimated in the equatorial oceans and at high latitudes in the North Pacific and Southern Oceans. The overestimation may be caused by limitation in plankton growth by iron, while underestimation in the sub-tropics partly stems from neglecting mesoscale variability.

The remaining discrepancies might be attributed in part to more complex processes involving nitrogen fixation. Finally, the tight coupling between carbon and either nitrogen or phosphate, which is generally implicit in ocean carbon models, precludes the simulation of past or future marine biological feedback mechanisms that involve a partial decoupling between carbon and nutrients.

Uptake of Anthropogenic CO_2 by the Ocean

Ocean uptake is constrained to some degree by observations of anthropogenic tracers. Three transient tracers are commonly used.

First, anthropogenic CO_2 itself gives a direct benchmark for model estimates of the quantity and distribution of anthropogenic CO_2 that has penetrated the ocean since the pre-industrial era.

Anthropogenic CO_2 can be inventoried by an indirect method whereby carbon concentration is compared to what would be expected from water exposed to pre-industrial air.

The $^{14}CO_2$ released in the early 1960s by atmospheric nuclear testing (commonly called bomb ^{14}C) provides a second tracer; the content of bomb ^{14}C in the ocean is used to constrain global air-sea CO_2 exchange, and ocean model results can be compared with its penetration depth as a benchmark for vertical transport.

Bomb ^{14}C is computed by subtracting the observed ^{14}C concentration from an estimate of its pre-industrial value.

Finally, CFCs also constrain the downward transport of tracers in ocean models. No natural background needs to be subtracted from CFCs.

None of these three tracers provide a perfect indicator of anthropogenic CO_2 uptake: CO_2 equilibrates with the atmosphere ten times faster than ^{14}C and ten times slower than CFCs; anthropogenic CO_2 and ^{14}C are indirectly estimated.

As part of the Ocean Carbon-Cycle Model Inter-comparison Project (OCMIP), a comparison of carbon models with respect to all three anthropogenic tracers is in progress.

Although regional estimates show discrepancies, modelled estimates of anthropogenic tracers agree reasonably well with observations when integrated globally.

The mean value of the penetration depth of bomb ^{14}C for all observational sites during the late 1970s is 390 ± 39 m. For the same years and stations, modelled estimates range between

283 and 376 m. Modelled and observed CFC concentrations have been compared locally but not yet globally.

Modelled anthropogenic CO_2 inventory since 1800 is comparable to the estimate of 40 ± 9 PgC for the Atlantic Ocean and 20 ± 3 PgC for the Indian Ocean. Latitude-depth profiles of anthropogenic CO_2 in the Atlantic, extracted from data and from models.

Modelled CO_2 uptake for the global ocean between 1800 and 1990 ranges between 100 and 133 PgC, comparable to the preliminary data-based estimate of 107 ± 27 PgC for the global ocean, which includes the Pacific value of 45 ± 15 PgC.

Although in reasonable agreement with basin and global estimates of anthropogenic CO_2, modelled inventories exhibit large differences at the regional scale: models tend to underestimate the inventory of anthropogenic CO_2 between 50°S and 50°N in the Atlantic and Indian Oceans, and to overestimate it at high latitudes.

In the Southern Ocean the uptake of anthropogenic CO_2 varies by a factor of two among models. The difficulty for models in reproducing the spatial structure of anthropogenic tracers may be indicative of problems in ocean physics mentioned earlier, and may be responsible for the increasing range of model estimates when future CO_2 uptake is projected by the same models.

The most recent model estimates of the ocean-atmosphere flux obtained with process-based models are 1.5 to 2.2 for 1980 to 1989, in agreement with earlier model estimates for the same period.

These estimates are fully consistent with the budget based on atmospheric observations alone, with estimates based on pCO_2 and ^{13}C observations, and with the SAR estimate of 2.0 ± 0.8 PgC/yr. The oceanic regions absorbing the largest quantities of anthropogenic CO_2 according to models are those where older waters come in contact with the atmosphere, such as high latitudes and upwelling regions of the equator.

In contrast, modelled sub-tropical regions rapidly saturate at atmospheric CO_2 level and do not absorb large quantities of anthropogenic CO_2.

Projections of CO_2 Concentration and their Implications

Terrestrial Carbon Model Responses to Scenarios of Change in CO_2 and Climate

Possible feedbacks from terrestrial carbon cycling to atmospheric CO_2 were assessed using multiple models by Cramer *et al.* Six DGVMs were driven first by CO_2 concentrations derived from the IS92a emissions scenario as in the SAR, and then with CO_2 changes plus climate changes derived from the HadCM$_2$ coupled ocean-atmosphere general circulation model simulation including sulphate aerosol effects as described by Mitchell *et al.*

Except for one empirical model, the models included explicit representation of all the following processes: the CO_2 fertilisation effect on NPP (modelled explicitly in terms of photosynthesis, respiration, and feedbacks associated with carbon allocation); responses of NPP to climate specific to each plant functional type (PFT); competition among PFTs for light and water; dynamic shifts in vegetation structure due to climate and CO_2 effects; competitive limits to above-ground biomass; natural disturbance regimes and their interaction with PFT composition; soil temperature and moisture effects on heterotrophic respiration.

Two models include an interactive N cycle. Land use and anthropogenic N deposition were not considered. Driven by increases in CO_2 beyond the present day, the modelled sink due to CO_2 fertilisation continued to increase.

By the middle of the 21st century the simulated land-atmosphere flux due to CO_2 was in the range -8.7 to -3.6 PgC/yr. Beyond mid-century the rate of increase became less, due to the declining photosynthetic response to CO_2.

When the climate change scenario was included as well as the CO_2 increase, modelled uptake was reduced compared with the CO_2-only analysis.

At mid-century, climate change reduced the uptake by 21 to 43%. A marked decline in terrestrial uptake after the midcentury was seen in two models, and one model had zero terrestrial uptake by 2100. By 2100 the range of model estimates of the land-atmosphere flux had widened to -6.7 to +0.4 PgC/yr. Increasing heterotrophic respiration in response to warming was a common factor (but not the only one) leading to reduced land uptake.

The differences among the modelled climate responses were largely due to unresolved discrepancies in the response of global NPP to temperature.

The balance of positive versus negative regional effects of climate change on NPP was estimated differently by these models, to the extent that the sign of the global response of NPP to climate change alone was not consistent. In addition, one model simulated a partial replacement of the Amazon rainforest by C_4 grassland.

This response was not shown, or occurred on a much smaller scale, in the other models. The details of this modelling exercise are presumably dependent on sensitivity of the particular climate model, and regional aspects of the simulated climate change.

Ocean Carbon Model Responses to Scenarios of Change in CO_2 and Climate

Analogous simulations have been performed with several ocean carbon models. To compute the impact of increasing CO_2 alone (no climate change), OCMIP models were forced to follow the atmospheric CO_2 concentration derived from the IS92a scenario as in the DGVM experiment.

All models agreed in projecting that the annual ocean-atmosphere flux of CO_2 continues to become larger, reaching -6.7 to -4.5 PgC/yr by 2100.

Since surface conditions (temperature, wind speed, alkalinity) were prescribed, the range in model estimates stems only from different representations of physical transport processes.

Several atmosphere-ocean models were used to project the effect of climate change. These models include most processes previously discussed, including all processes associated with carbonate chemistry and gas exchange, physical and biological transport of CO_2, and changes in temperature, salinity, wind speed, and ice cover.

They account for simple changes in biological productivity, but not for changes in external nutrient supply, species composition, pH, or Redfield ratios, all of which could be involved in more complex biological feedbacks.

Coupled models estimate the impact of climate change as a departure, reported in per cent, from a "control" experiment modelling the effect of increasing atmospheric CO_2 alone.

In climate change simulations, warming of surface waters and increased stratification of the upper ocean produced an overall positive feedback that reduced the accumulated ocean uptake of CO_2 by 6 to 25% between 1990 and the middle of the 21st century, as compared with the CO_2-only case. In the first part of the simulation, the climate-mediated feedback is mainly due to the temperature effect on CO_2 solubility.

Towards the mid-century, the impact of circulation changes becomes significant in most models, with the net effect of further reducing ocean CO_2 uptake.

To investigate the effect of climate change on the IS92a scenario, the average of the OCMIP CO_2-only projections was used as a baseline and the reduction in atmosphere-ocean CO_2 flux caused by climate change (in per cent since the beginning of the simulation) was applied to this curve.

The range in model results must be attributed to uncertainties related to climate change feedback, and not to uncertainties in the modelling of physical transport. The range of model estimates of the climate change impact is dependent on the choice of scenario for atmospheric CO_2 and on assumptions concerning marine biology.

At high CO_2 concentrations, marine biology can have a greater impact on atmospheric CO_2 than at low concentrations because the buffering capacity of the ocean is reduced.

Although the impact of changes in marine biology is highly uncertain and many key processes are not included in current models, sensitivity studies can provide approximate upper and lower bounds for the potential impact of marine biology on future ocean CO_2 uptake.

A sensitivity study of two extreme scenarios for nutrient supply to marine biology gave a range of 8 to 25% for the reduction of CO_2 uptake by mid-century. This range is comparable to other uncertainties, including those stemming from physical transport.

Coupled Model Responses and Implications for Future CO_2 Concentrations

Carbon cycle models have indicated the potential for climate change to influence the rate of CO_2 uptake by both land and oceans and thereby influence the time course of atmospheric CO_2 concentration for any given emissions scenario. Coupled models are required to quantify these effects.

Two general circulation model simulations have included interactive land and ocean carbon cycle components. The Cox *et al.* model was driven by CO_2 emissions from the IS92a scenario and the Friedlingstein *et al.* model was driven by CO_2 emissions from the SRES A2 scenario. Both simulations indicate a positive feedback, i.e., both CO_2 concentrations and climate change at the end of the 21st century are increased due to the coupling.

The simulated magnitudes of the effect differ (+70 ppm). In the Cox *et al.* simulation, which included a DGVM, the increased atmospheric CO_2 is caused mainly by loss of soil carbon and in part by tropical forest die back.

The magnitude of the climate-carbon cycle feedback still has large uncertainties associated with the response of the terrestrial biosphere to climate change, especially the response of heterotrophic respiration and tropical forest NPP to temperature. In the following section, simplified models are used to assess these uncertainties.

Methods for Assessing the Response of Atmospheric CO_2 to Different Emissions Pathways and Model Sensitivities

This section follows the approach of previous IPCC

reports in using simplified, fast models (sometimes known as reduced-form models) to assess the relationship between CO_2 emissions and concentrations, under various assumptions about their future time course. The models lend themselves to somewhat different approaches to estimating uncertainties. In the ISAM model, "high-CO_2" and "low-CO_2" alternatives are calculated for every emissions scenario, based on tuning the model to match the range of responses included in the model intercomparisons. Uncertainties cited from the ISAM model can be regarded as providing a lower bound on uncertainty since they do not admit possible behaviours outside the range considered in recent modelling studies.

In the Bern-CC model, "high-CO_2" and "low-CO_2" alternatives are calculated by making bounding assumptions about carbon cycle processes (for example, in the high-CO_2 parametrization CO_2 fertilisation is capped at year 2000; in the low-CO_2 parametrization Rh does not increase with warming).

This approach yields generally larger ranges of projected CO_2 concentrations than the ISAM approach. The ranges cited from the Bern-CC model can be regarded as approaching an upper bound on uncertainty, since the true system response is likely to be less extreme than the bounding assumptions, and because the combination of "best" and "worst" case assumptions for every process is intrinsically unlikely.

Concentration Projections Based on IS92a, for Comparison with Previous Studies

Illustrative model runs based on the IS92a scenario are shown first so as to allow comparison with earlier model results presented in the SAR and the SRRF. In the SRRF comparison of eighteen global carbon cycle models the CO_2 fertilisation response of the land was calibrated to match the central estimate of the global carbon budget for the 1980s, assuming a land-use source of 1.6 PgC/yr in the 1980s and attributing the residual terrestrial sink to CO_2 fertilisation. This intercomparison yielded CO_2 concentrations in 2100 of 668 to 734 ppm; results presented in Schimel *et al.* (from the Bern model) gave 688 ppm.

After recalibrating to match a presumed land-use source of 1.1 PgC/yr, implying a weaker CO_2 response, the 2100 CO_2 concentration was given as 712 ppm in the SAR. An IPCC Technical Paper evaluated the sensitivity of IS92a results to this calibration procedure.

Wigley *et al.*, found that a range of assumed values from 0.4 to 1.8 PgC/yr for the land-use source during the 1980s gave rise to a range of 2100 CO_2 concentrations from 667 to 766 ppm. In contrast with the SAR, the results presented here are based on approximating the behaviour of spatially resolved processbased models in which CO_2 and climate responses are not constrained by prior assumptions about the global carbon budget.

The CO_2-only response of both models' reference cases leads to a 2100 CO_2 concentration of 682 ppm (ISAM) or 651 ppm (Bern-CC). These values are slightly lower than projected in the SAR -715 ppm (ISAM; SAR version) and 712 ppm (Bern Model; SAR version) - because current process-based terrestrial models typically yield a stronger CO_2 response than was assumed in the SAR.

With climate feedbacks included, the 2100 CO_2 concentration in the reference case becomes, by coincidence, effectively indistinguishable from that given in the SAR: 723 ppm (ISAM) and 706 ppm (Bern-CC).

The ranges of 164 ppm or -12%/+11% (about the reference case) (ISAM) and 273 ppm or -10%/+28% (Bern-CC) in the 2100 CO_2 concentration indicate that there is significant uncertainty about the future CO_2 concentrations due to any one pathway of changes in emissions. Separate calculations with the Bern-CC model show that the effect of changing climate sensitivity alone is less important than the effect of varying assumptions in the carbon cycle model's components.

The effect of increasing climate sensitivity to 4.5 °C (increasing the climate feedback) is much larger than the effect of reducing climate sensitivity to 1.5 °C. The "low-CO_2" parametrization of Bern-CC yields CO_2 concentrations closer to the reference case than the "high-CO_2" parametrization, in which the terrestrial sink is forced to approach zero during the first few decades of the century due to the capping of the CO_2 fertilisation effect.

The reference simulations with ISAM yielded an implied average land-use source during the 1980s of 0.9 PgC/yr. The range was 0.2 to 2.0 PgC/yr. Corresponding values for Bern-CC were 0.6 PgC/yr and a range of 0.0 to 1.5 PgC/yr. These ranges broadly overlap the range estimates of the 1980s land-use source. Present knowledge of the carbon budget is therefore not precise enough to allow much narrowing of the uncertainty associated with future land and ocean uptake as expressed in these projections.

SRES Scenarios and their Implications for Future CO_2 Concentration

The Special Report on Emissions Scenarios (SRES) produced a series of scenarios, of which six are used here, representing outcomes of distinct narratives of economic development and demographic and technological change. In ISAM model runs with these scenarios, past fossil emissions, CO_2 concentrations and mean global temperatures were specified up to and including 1999; scenario-based analyses started in 2000.

In the Bern-CC model runs, observed CO_2 and past fossil emissions were prescribed, and historical temperature changes were modelled, based on radiative forcing from greenhouse gases and aerosols; again, scenario-based analyses started in 2000. Past emissions from changing land use were calculated in order to balance the carbon budget.

The six scenarios lead to substantial differences in projected CO_2 concentration trajectories. Significant uncertainties are introduced by the range of model parametrizations considered, so that the trajectories calculated for "adjacent" scenarios overlap, especially during the first half-century.

The reference cases of the six scenarios account for a range of 2100 CO_2 concentrations from 541 to 963 ppm in the Bern-CC model and 549 to 970 ppm in the ISAM model. The uncertainties around the 2100 values due to model parametrizations are -12 to +10 % (ISAM) and -14 to +31 % (Bern-CC).

These uncertainties reflect incomplete understanding of climate sensitivity and the carbon cycle. They substantially limit our current ability to make quantitative predictions about the future consequences of a given emissions trajectory. Nevertheless, the results show that higher emissions are always expected to lead to higher projected atmospheric concentrations. They also show that the range of emissions scenarios currently accepted as plausible leads to a range of CO_2 concentrations that exceeds the likely upper bound of uncertainties due to differences among model parameterizations and assumptions.

Stabilisation Scenarios and their Implications for Future CO_2 Emissions

Stabilisation scenarios illustrate implied rates of CO_2 emission that would arrive at various stable CO_2 concentration levels.

These have been projected using a similar methodology to that applied in the analysis of emissions scenarios. The WRE trajectories follow CO_2 concentrations consistent with the IS92a scenario beginning in 1990 and branch off to reach constant CO_2 concentrations of 450, 550, 650, 750 and 1,000 ppm.

The rationale for various alternative time trajectories and stabilisation levels is discussed in earlier Chapter of the IPCC WGIII Third Assessment Report. Differences in emissions pathways for different time trajectories leading to a certain stabilisation target (e.g., S versus WRE profiles) are discussed in Schimel *et al.*

Here, we have calculated emissions for one set of emission profiles to illustrate differences in implied emissions that arise from updating models since the SAR. Then anthropogenic emissions for the prescribed CO_2 stabilization profiles were calculated; deduced emissions equal the change in modeled ocean and terrestrial carbon inventories plus the prescribed change in atmospheric CO_2 content. To estimate the strength of carbon cycle-climate feedbacks, global temperature (ISAM) and changes in the fields of temperature, precipitation and cloud cover (Bern-CC) were projected from CO_2 radiative forcing only, neglecting effects of other greenhouse gases and aerosols which are not specified in the WRE profiles. The results for the reference cases are not substantially different from those presented in the SAR.

However, the range based on alternative model parametrizations is larger than presented in the SAR, mainly due to the range of simulated terrestrial CO_2 uptake. CO_2 stabilisation at 450, 650 or 1,000ppm would require global anthropogenic CO_2 emissions to drop below 1990 levels, within a few decades, about a century, or about two centuries, respectively.

In all cases, once CO_2 concentration becomes constant, the implied anthropogenic emission declines steadily. This result was expected. It highlights the fact that to maintain a constant future CO_2 concentration, anthropogenic CO_2 emissions would ultimately have to be reduced to the level of persistent natural sinks.

Persistent terrestrial sinks are not well quantified; peatlands may be a candidate, but the gradual rise in atmospheric CO_2 concentration during the present interglacial argues against any such sink. Estimates of current uptake by peatlands are <0.1 PgC/yr. Mixing of ocean DIC between surface and deep waters should continue to produce ocean uptake for several centuries after an input of anthropogenic atmospheric CO_2. This mixing is the main reason for continued uptake (and therefore positive calculated emissions) after stabilisation.

However, the main, known natural sink expected to persist longer than a few centuries is that due to dissolution of $CaCO_3$ in ocean sediments, which increases ocean alkalinity and thereby allows additional CO_2 to dissolve in the ocean. For CO_2 concentrations about 1,000 ppm, this sink is estimated to be smaller than about -0.1 PgC/yr. Thus, for any significant CO_2 emissions to persist over centuries without continuing to increase atmospheric CO_2 would require some method of producing an artificial carbon sink.

The differences among the CO_2 concentrations projected with the various SRES scenarios considered are larger than the differences caused by inclusion or omission of climate-mediated feedbacks. The range of uptake rates projected by process-based models for any one scenario is, however, considerable, due to uncertainties about (especially) terrestrial ecosystem responses to high CO_2 concentrations, which have not yet been resolved experimentally, and uncertainties about the response of global NPP to changes in climate. A smaller feedback would be implied if, as some models indicate, global NPP increases with warming throughout the relevant range of climates and no forest die back occurs.

Larger positive feedbacks would be implied if regional drying caused partial die back of tropical forests, as some of the DGVMs in Cramer *et al*, and one coupled climatecarbon model study of Cox *et al*, suggest; however, another coupled climate-carbon model study suggests a smaller feedback. Uncertainty also arises due to differences in the climate responses of ocean models, especially as regards the extent and effects (biological as well as physical) of increased stratification in a warmer climate.

In conclusion, *anthropogenic CO_2 emissions are virtually certain to be the dominant factor determining CO_2 concentrations throughout the 21st century*. The importance of anthropogenic emissions is underlined by the expectation that the proportion of emissions taken up by both ocean and land will decline at high atmospheric CO_2 concentrations (even if absolute uptake by the ocean continues to rise).

There is considerable uncertainty in projections of future CO_2 concentration, because of uncertainty about the effects of climate change on the processes determining ocean and land uptake of CO_2. These uncertainties do not negate the main finding that anthropogenic emissions will be the main control.

Large-scale manipulations of terrestrial ecosystems have been proposed as a means of slowing the increase of atmospheric CO_2 during the 21st century in support of the aims of the Kyoto Protocol. Based on current understanding of land use in the carbon cycle, the impacts of future land use on terrestrial biosphere-atmosphere exchanges have the potential to modify atmospheric CO_2 concentrations on this time-scale.

Direct effects of land-use changes are thought to represent about 10 to 30% of total anthropogenic CO_2 emissions, so there is scope for either intended or unintended changes in land use to reduce or increase total anthropogenic emissions. But the possibilities for enhancing natural sinks have to be placed in perspective: a rough upper bound for the reduction in CO_2 concentration that could be achieved by enhancing terrestrial carbon uptake through land-use change over the coming century is 40 to 70 ppm, to be considered against a two to four times larger potential for increasing CO_2 concentraion by deforestation, and a >400 ppm range among the SRES scenarios.

Index

D

E

F

G

H

O

P

R

S

T

U

V

W